Computational Intelligence in the Industry 4.0

This book discusses the importance of using industrial intelligence in collaboration with computational intelligence in forming a smart system for diverse applications. It further illustrates the challenges and deployment issues in industrial resolution. The text highlights innovation and applications of computational agents and the industrial intelligence era to automate the requirements as per Industry 4.0.

This book:

- Discusses computational agents for handling automation issues and the role of ethics in industrial resolution.
- Presents intelligence approaches for products, operations, systems, and services.
- Illustrates the fundamentals of computational intelligence to forecast and analyze the requirements of society for automation as well as recent innovations and applications.
- Highlights computation intelligence approaches in reducing human effort and automating the analysis of the production unit.
- Showcases current innovation and applications of computational agents and industrial intelligence as per Industry 4.0.

The text is primarily written for senior undergraduate and graduate students, and academic researchers in diverse fields including electrical engineering, electronics and communication engineering, industrial engineering, manufacturing engineering, and computer science engineering.

Intelligent Data-Driven Systems and Artificial Intelligence
Series Editor: Harish Garg

Modelling of Virtual Worlds Using the Internet of Things
Edited by Simar Preet Singh and Arun Solanki

Data-Driven Technologies and Artificial Intelligence in Supply Chain Tools and Techniques
Mahesh Chand, Vineet Jain, and Puneeta Ajmera

For more information about this series, please visit: www.routledge.com/ Intelligent-Data-Driven-Systems-and-Artificial-Intelligence/book-series/ CRCIDDSAAI

Computational Intelligence in the Industry 4.0

Edited by
Anil Kumar Dubey, Vikash Yadav,
and Munesh Chandra Trivedi

CRC Press
Taylor & Francis Group
Boca Raton London New York

CRC Press is an imprint of the
Taylor & Francis Group, an **informa** business

Front cover image: PSboom/Shutterstock

First edition published 2024
by CRC Press
2385 NW Executive Center Drive, Suite 320, Boca Raton FL 33431

and by CRC Press
4 Park Square, Milton Park, Abingdon, Oxon, OX14 4RN

CRC Press is an imprint of Taylor & Francis Group, LLC

ISBN: 978-1-032-54056-6 (hbk)
ISBN: 978-1-032-76557-0 (pbk)
ISBN: 978-1-003-47903-1 (ebk)

DOI: 10.1201/9781003479031

Typeset in Sabon
by codeMantra

Contents

Preface

The Internet of Things is an innovative technology to facilitate the connection of abundant devices through the internet and monitor them. The fundamentals of industrial IoT along with tools and techniques aim to establish infrastructure that reduces human effort and automates the industry system for production, development, and processing. Some sensors are used to establish the required infrastructure and protocols to assist them for automation process.

Chapter 1 lays the foundation of automation with important types of automation such as Industrial Automation, Home Automation, Process Automation, Robotic Process Automation (RPA), IT Automation, and Test Automation.

Chapter 2 covers RPA and test automation in which Software Bots, User Interface Interaction, Rule-Based Automation, Unattended and Attended Automation, and Workflow Automation are discussed.

Chapter 3 explores the intersection of intelligent computing and cloud computing, highlighting the benefits and challenges of combining these two fields. Intelligent computing techniques including machine learning, deep learning, natural language processing, and data analytics can be used to optimize cloud computing operations, improve resource allocation, and enhance the security and privacy of cloud-based applications.

Chapter 4 expands several popular industrial automation communication protocols such as Modbus, Profibus, DeviceNet, Ethernet/IP, Profinet, CAN, Foundation Fieldbus, HART, Modbus TCP/IP, and OPC.

Chapter 5 discusses MarketWatch and creates a complete Mobile+ Web application that will enable users to create portfolios, buy and sell stocks using virtual currency, and experience real-time market activity without actually losing any money from existing literature.

Owing to the increase in cyber threats such as hacking, phishing, virus assaults, and identity theft, cybersecurity has developed over time, which is discussed in Chapter 6. In this chapter, we also explore the need to study cybersecurity with its trending challenges, emerging trends, and threats related to automation.

Chapter 7 focuses on intelligence approaches for products, operations, systems, and service. As machine learning and data science play a vital role in automation and contribute to the understanding of various diseases, we focus on Parkinson's disease detection using machine learning models in Chapter 8.

Chapter 9 examines the key enablers of a comparison between nonlinear mapping and high-resolution image.

Chapter 10 discusses the key concepts of deployment issues in industrial resolution and typical deployment concern in industrial setting. The increasing demands of machine learning in each automation process motivate the study of Industry 4.0, so we focus on hate speech detection using machine learning models in Chapter 11. A crucial piece of Industry 4.0, or the fourth industrial revolution, technology is digital twins. They build simulations that can forecast a product or process and performance using real-world data. The simulation is based on both the assets and present state and previous data. With the use of digital twins, it is possible to increase system performance while keeping costs low, decreasing physical testing, and enhancing product quality.

In Chapter 12, digital twin technology for industrial automation is discussed.

Chapters 13 highlights the theoretical analysis of simple retrieval engine.

About the editors

Dr. Anil Kumar Dubey has more than 10 years of teaching experience, of which 6 years were as Associate Professor and 8 years post-PhD. Currently he is working as Associate Professor (CSE) at ABES Engineering College in Ghaziabad. Prior to this, he worked as Associate Professor (CSE) at Poornima Institute of Engineering & Technology, Jaipur. Dr. Dubey successfully filed 15 patents (12 national and 3 international patents), of which 6 patents have been granted. He has published several textbooks and research papers in different international journals and proceedings of repute. He has received numerous awards including Young Scientist, Best Paper, and Best Trainer. Dr. Dubey has organized more than 20 international conferences technically sponsored by IEEE, ACM, and Springer.

Dr. Vikash Yadav is currently working as a Computer Lecturer in the Department of Technical Education, Uttar Pradesh, India. He has completed his PhD in Computer Science and Engineering from Dr. A.P.J. Abdul Kalam Technical University (State Government University), Lucknow, Uttar Pradesh, India. He received his M.Tech. in Software Engineering from Motilal Nehru National Institute of Technology, Prayagraj, Allahabad, Uttar Pradesh, and B.Tech. in Computer Science and Engineering from U.P. Technical University, Lucknow, India. Dr. Yadav is a life member of the Computer Society of India (CSI) and also a life member of IAENG. His research interests include Data Mining, Image Processing, and Machine Learning. Prior to the current assignment, he has worked for ABES Engineering College, Ghaziabad. He is also the Editorial Board Member of the Journal *Recent Advances in Electrical & Electronic Engineering* (Scopus Indexed Journal), Bentham Science Publication. He has published nearly 70 research papers in various international journals (SCI/SCIE/Scopus) and conferences of repute. He has edited two books from CRC Press and Bentham Science Publication and two are currently in process; he has also edited several special issues for SCI/SCIE/Scopus journals. He also chaired several international conferences as a session chair. He has published four Indian Patents. He has more than 450 Google Scholar, 12 H-index, and 15 i-10 Index citations.

Dr. Munesh Chandra Trivedi has more than 20 years of teaching experience, of which 4 years was as Professor and 13 years post-PhD. Currently he is a Professor (CSE) at NCERT, Faculty of Engineering and Technology, PSSCIVE, Bhopal (an apex institution of the Ministry of Education, Government of India). Prior to this, he worked as Associate Professor (CSE), National Institute of Technology, Agartala, Tripura, India, and Associate Professor and HoD (IT), Rajkiya Engineering College, Azamgarh, Uttar Pradesh (State Government institution of Uttar Pradesh, India), with the additional responsibility of Dean of Academics and Associate Dean (UG Program), Dr. A.P.J. Abdul Kalam Technical University, Lucknow (State Technical University). Dr. Chandra was also the Director (In-Charge) of Rajkiya Engineering College, Azamgarh. He successfully filed 63 patents (54 national and 9 international patents in Germany, South Africa, and Australia), of which 27 patents have been granted. He has published 12 textbooks and 153 research papers in different international journals and proceedings of repute. Dr. Chandra edited 38 books for Springer Nature. He has successfully supervised 14 PhD students and received numerous awards including the Young Scientist Visiting Fellowship, Albert Einstein Research Scientist Award, Best Senior Faculty Award, Outstanding Scientist, Dronacharya Award, Author of Year, and Vigyan Ratan Award from different national and international forums. Dr. Chandra has organized more than 32 international conferences technically sponsored by IEEE, ACM, and Springer.

Contributors

Rohit Anand
G B Pant Engineering College, New Delhi, India Shashank Awasthi
GL Bajaj Institute of Technology and Management
Greater Noida, India

Richa Bansal
Bhagwant University
Rajasthan, India

Abhinav Dahiya
University Institute of Engineering & Technology
Maharshi Dayanand University
Haryana, India

Anil Kumar Dubey
ABES Engineering College
Ghaziabad, India

Mohammad Faiz
School of Computer Science and Engineering
Lovely Professional University
Phagwara, India

Pradeep Gupta
Ajay Kumar Garg Engineering College
Ghaziabad, India

Sonam Gupta
Ajay Kumar Garg Engineering College
Ghaziabad, India

Rabban Javed
AIIT, Amity University Noida, Uttar Pradesh, India

Nishant Jawla
KIET Group of Institutions
Ghaziabad, India

Kamaldeep Joshi
University Institute of Engineering & Technology
Maharshi Dayanand University
Haryana, India

Vimal Kumar
School of Computer Science and Engineering
Galgotias University
Greater Noida, India

Chuan-Ming Liu
National Taipei University of Technology (Taipei Tech)
Taipei, Taiwan

Pawan Kumar Mall
GL Bajaj Institute of Technology
  and Management
Greater Noida, India

Anurag Mishra
KIET Group of Institutions
Ghaziabad, India

Vipul Narayan
School of Computer Science and
  Engineering
Galgotias University
Greater Noida, India

Pranjal Rai
KIET Group of Institutions
Ghaziabad, India

Ravi Saini
Department of Computer Science
Government College for Women
  Gurawara
Haryana, India

Mala Saraswat
Bennett University
Greater Noida, India

Anjali Sharma
Amity Institute of Information
  Technology (AIIT)
Noida, India

Satyam Sharma
KIET Group of Institutions
Ghaziabad, India

Abhishek Kumar Shukla
ABES Engineering College
Ghaziabad, India

Nidhi Sindhwani
Amity Institute of Information
  Technology (AIIT)
Noida, India

Swapnita Srivastava
GL Bajaj Institute of Technology
  and Management
Greater Noida, India

Erma Suryani
Institut Teknologi Sepuluh
  Nopember (ITS)
Surabaya, Indonesia

Rashmi Vashisth
Amity Institute of Information
  Technology (AIIT)
Noida, India

Vikash Yadav
Government Polytechnic Bighapur
  Unnao, Department of Technical
  Education,
Uttar Pradesh, India

Chapter 1

Introduction to automation

Mala Saraswat and Anil Kumar Dubey

Automation is the process of using technology, equipment, and software to carry out operations with little or no human involvement. Automation aims to increase productivity, accuracy, speed, and dependability across a range of businesses and activities. It entails swapping out manual, time-consuming, repetitive processes with automated ones that can be completed reliably and without continuous human oversight (Table 1.1).

Numerous industries, including manufacturing, agriculture, finance, healthcare, transportation, and more, can benefit from automation. It includes both mechanical and robotic processes that take place in the real world and digital processes that deal with data and information using software and algorithms.

MAJOR BENEFITS OF AUTOMATION

Although automation has many advantages, it's vital to think about how technology can affect employment because it might make some positions obsolete. Finding a balance between automation and the human worker is increasingly important as technology develops (Figure 1.1).

INDUSTRIAL AUTOMATION

The use of cutting-edge technology and control systems to automate numerous jobs, processes, and operations in manufacturing and industrial settings is known as industrial automation [1–11]. Industrial automation aims to decrease the need for manual intervention while increasing efficiency, productivity, safety, and quality. To construct a system that works flawlessly, several pieces of hardware and software are integrated.

DOI: 10.1201/9781003479031-1

1

Table 1.1 Important automation types

Types of automation	Description
Industrial automation	This type involves using robotic systems and equipment to automate manufacturing procedures and tasks. It may involve activities including packaging, material handling, assembly, and quality control.
Home automation	Automation of numerous household functions, including lighting, heating, security systems, entertainment systems, and appliances, is what is referred to as "smart home technology." It strives to make homes more practical, secure, and energy-efficient.
Process automation	This refers to employing software and digital tools to automate routine and rule-based business activities. It is frequently used to streamline operations and lower human error in industries including banking, human resources, and customer support.
Robotic process automation (RPA)	RPA automates processes within software applications by using software robots (bots). These bots are helpful for activities like data entry, data extraction, and report preparation because they can simulate how people interact with computers.
IT automation	Involves automating a variety of IT-related operations, including network configuration, server provisioning, software deployment, and system monitoring. IT teams can better manage challenging settings because of it.
Test automation [21-24]	To automate the testing of software applications, this is frequently used in software development. Automated testing systems can swiftly and accurately carry out repetitive tests, ensuring software quality and finding problems.

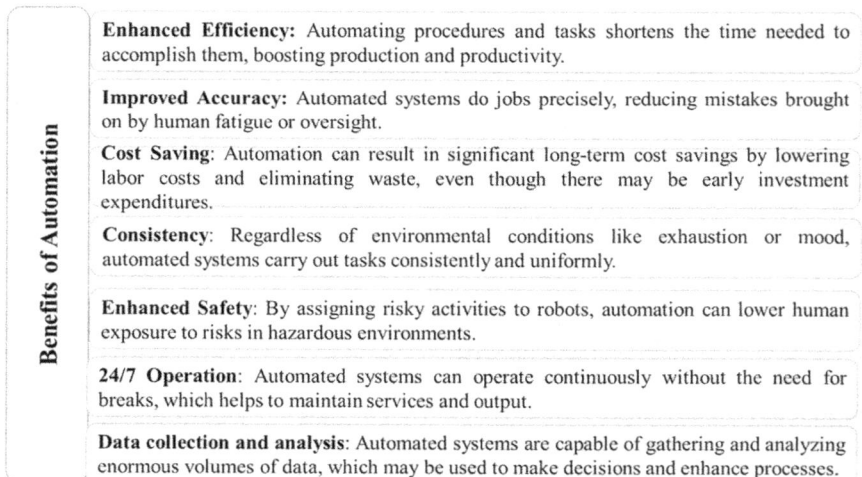

Benefits of Automation

Enhanced Efficiency: Automating procedures and tasks shortens the time needed to accomplish them, boosting production and productivity.

Improved Accuracy: Automated systems do jobs precisely, reducing mistakes brought on by human fatigue or oversight.

Cost Saving: Automation can result in significant long-term cost savings by lowering labor costs and eliminating waste, even though there may be early investment expenditures.

Consistency: Regardless of environmental conditions like exhaustion or mood, automated systems carry out tasks consistently and uniformly.

Enhanced Safety: By assigning risky activities to robots, automation can lower human exposure to risks in hazardous environments.

24/7 Operation: Automated systems can operate continuously without the need for breaks, which helps to maintain services and output.

Data collection and analysis: Automated systems are capable of gathering and analyzing enormous volumes of data, which may be used to make decisions and enhance processes.

Figure 1.1 Major benefits of automation.

Key components of industrial automation

- **Sensors and Actuators:** Sensors are tools that gather information about the physical environment, such as position, pressure, and temperature. Robotic arms can move or valves can open and close with the help of actuators, which are devices that carry out operations depending on data from sensors.
- **Programmable Logic Controllers (PLCs):** PLCs are specialist computers that manage and keep an eye on real-time operations for equipment and processes. They take in data from sensors, process it, and then provide instructions to actuators. Many industrial automation systems are based on PLCs.
- **Human-Machine Interface (HMI):** HMIs give engineers and operators a graphical interface through which to communicate with the automation system. They provide real-time information, alarms, and control adjustments.
- **Supervisory Control and Data Acquisition (SCADA) Systems:** SCADA systems are employed to remotely monitor and manage industrial activities. They gather information from numerous sensors and gadgets and empower operators to make wise decisions and corrections.
- **Industrial Robots:** Robotic systems are employed for a variety of operations, including assembling, welding, painting, and material handling. These robots can operate alongside people in cooperative contexts and can do repetitive jobs with high precision.
- **Motion Control Systems:** By regulating the movement of machinery and equipment, these systems guarantee accurate positioning, rapid movement, and synchronization. They are frequently employed in manufacturing procedures that call for precise motion, like CNC machines.
- **Automated Guided Vehicles (AGVs):** Mobile robots called AGVs are used to move items around production plants. They are applied to automation of logistics and material handling tasks.

Despite the fact that industrial automation has many benefits, it is important to take into account aspects like the initial investment expenses, system upkeep, and any potential labor effects. Maximizing the advantages of industrial automation while addressing possible issues requires careful design and integration (Table 1.2).

Sensors and actuators

Fundamental elements of many automation systems, from industrial processes and robotics to smart home technologies and IoT (Internet of Things) applications, are sensors and actuators. They are essential for

Table 1.2 Benefits of industrial automation

Benefits	Description
Increased productivity	Automation enables processes to be run continuously without interruption, increasing productivity and throughput.
Improved quality	Automated production methods can reliably manufacture goods with a high degree of accuracy and uniformity, minimizing flaws and inconsistencies.
Reduced labor costs	For tedious and taxing activities, automation can take the place of manual labor, saving money over time.
Enhanced safety	Automated systems can be used for dangerous operations, lowering the risk of mishaps and harm to human workers.
Flexibility	To adjust to changes in production requirements or product designs, automation systems can be reprogrammed or changed.
Data collection and analysis	Massive amounts of data are produced by automation systems, which can be evaluated to spot inefficiencies and places for development.
Shorter cycle times	Automation can result in shorter production cycles, allowing businesses to respond to client requests more rapidly.
Consistency	Automation reduces the variability in product quality by performing jobs with a high degree of consistency and dependability.

gathering environmental data, allowing machines to communicate with their environments, and enabling the automation of numerous processes and jobs.

Sensors: Devices known as sensors are used to identify and gauge physical parameters or environmental conditions. They transform physical events like temperature, pressure, light, motion, or chemical characteristics into electrical signals that automation systems can process and decipher. Automation [20] is able to adapt to changes in the environment because of the crucial input that sensors provide to control systems and algorithms (Figure 1.2).

Actuators: In reaction to control signals from automation systems, actuators are devices that carry out physical operations. They enable machines to engage with the physical world by converting electrical or digital signals into mechanical action (Figure 1.3).

Automation closed-loop control systems are made possible by the cooperation of sensors and actuators. Controllers process the data that sensors offer about the environment or the state of a system. To accomplish certain tasks, the controllers produce control signals that are transmitted to the actuators. This feedback loop makes sure that automation systems are capable of making wise decisions and completing tasks with accuracy and precision.

Temperature Sensors: Utilize resistance temperature detectors (RTDs) or thermocouples to measure temperature fluctuations.

Pressure Sensors: Uses for these sensors include industrial monitoring and control, automotive systems, and medical equipment. They also detect changes in pressure.

Proximity Sensors: Determine the presence or absence of an object without physical contact. Common types include capacitive, inductive, and ultrasonic sensors.

Motion Sensors: Detect movement or changes in position and are widely used in security systems, robotics, and gaming devices.

Light Sensors: Determine the brightness or ambient light levels. Examples include photodiodes, phototransistors, and light-dependent resistors (LDRs).

Humidity Sensors: For environmental control and climate monitoring, determine the relative humidity in the air.

Gas Sensors: Identification of specific gases in the environment is essential for safety and pollution prevention.

Important Sensor

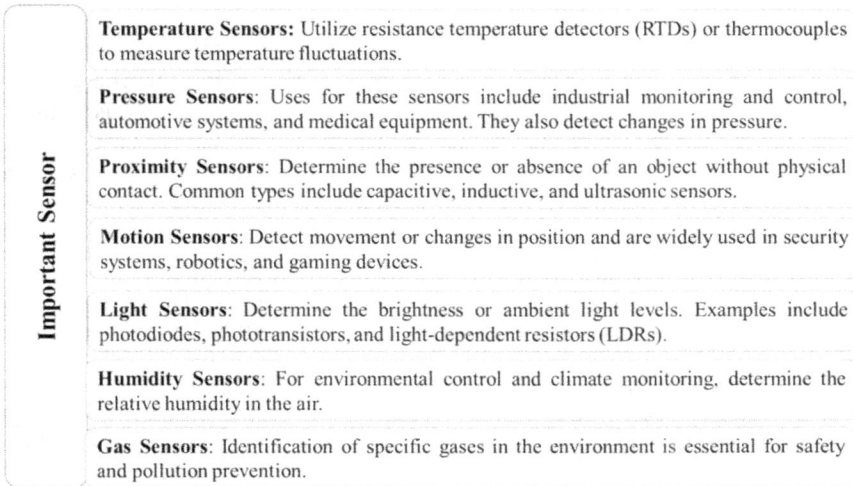

Figure 1.2 Some important sensors.

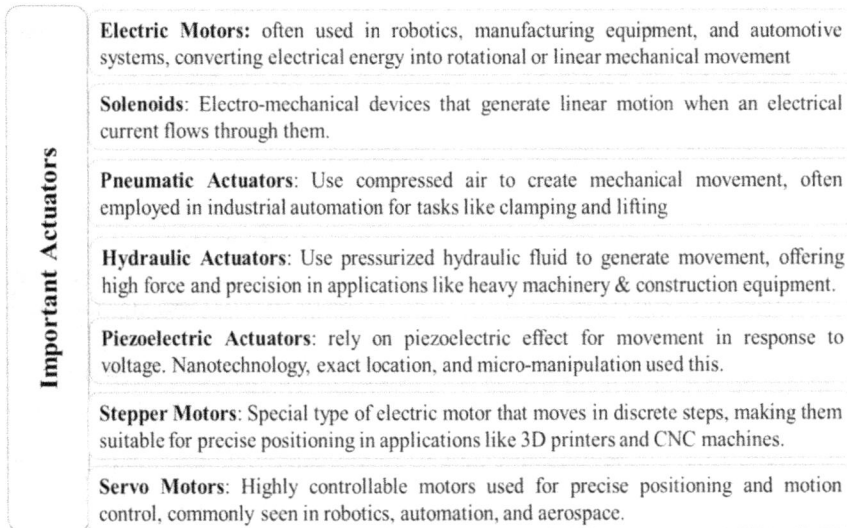

Electric Motors: often used in robotics, manufacturing equipment, and automotive systems, converting electrical energy into rotational or linear mechanical movement

Solenoids: Electro-mechanical devices that generate linear motion when an electrical current flows through them.

Pneumatic Actuators: Use compressed air to create mechanical movement, often employed in industrial automation for tasks like clamping and lifting

Hydraulic Actuators: Use pressurized hydraulic fluid to generate movement, offering high force and precision in applications like heavy machinery & construction equipment.

Piezoelectric Actuators: rely on piezoelectric effect for movement in response to voltage. Nanotechnology, exact location, and micro-manipulation used this.

Stepper Motors: Special type of electric motor that moves in discrete steps, making them suitable for precise positioning in applications like 3D printers and CNC machines.

Servo Motors: Highly controllable motors used for precise positioning and motion control, commonly seen in robotics, automation, and aerospace.

Important Actuators

Figure 1.3 Important actuators.

Programmable logic controllers

PLCs are specialized digital computers that are used in industrial automation and control systems to monitor and manage numerous operations, processes, and machines. PLCs can execute programmed logic to automate complex activities and processes and are made to resist harsh industrial conditions.

The essential elements of PLCs

- **Input/Output Modules (I/O Modules):** These modules interface with sensors, actuators, and other devices to link the PLC to the outside world. They give the PLC a way to deliver outputs, such as control signals to actuators, and receive inputs, such as sensor readings.
- **Central Processing Unit (CPU):** The "brain" or CPU of the PLC is in charge of carrying out pre-programmed logic. It processes input signals, runs the control program, and produces output signals in accordance with the logic of the program.
- **Memory:** Program memory, data memory, and retentive memory are several forms of memory found in PLCs. Program memory is used to store the control program, while data memory is used to store variables and intermediate values.
- **Programming Interface:** Engineers and programmers can build and edit the control logic for PLCs using specialized software. Ladder logic, structured text, function block diagrams, and other programming languages are frequently used for PLCs.
- **Communication Ports:** Modern PLCs frequently have communication connections for connecting to other devices, including computers, networks, and HMIs, enabling remote monitoring, control, and data exchange.

PLC operation and applications

- **Logic Control:** PLCs carry out the control logic that engineers have programmed. To regulate how machines and processes operate, this logic also uses timers, counters, and other instructions.
- **Sequence Control:** PLCs are frequently used to automate procedures that call for certain order of steps, such as manufacturing assembly lines, conveyor systems, and batch processing.
- **Feedback Control:** PLCs can use sensor feedback to maintain and change the process parameters that are needed, maintaining precision and consistency.
- **Safety Control:** By monitoring key conditions and initiating the necessary actions, PLCs can be configured to conduct safety measures, such as emergency shutdowns.
- **Data Logging and Reporting:** Process parameters, performance, and event data can be recorded by PLCs and used for analysis, optimization, and reporting.
- **Motion Control:** For accurate motion in applications like robotics and CNC machines, PLCs can operate servo motors, stepper motors, and other motion devices.
- **HMI Integration:** HMIs and PLCs frequently connect, enabling operators to observe operations, get alarms, and engage with the control system.

Benefits of PLCs

- **Reliability:** PLCs are made to operate robustly and dependably in industrial settings, reducing downtime.
- **Flexibility:** Without requiring large hardware changes, PLC programming can be quickly changed and adjusted to changes in production requirements.
- **Accuracy:** PLCs guarantee precise process synchronization and control, resulting in consistently high product quality.
- **Scalability:** To meet varying process capacities and complexity, PLC systems can be scaled up or down.
- **Maintenance:** PLCs make maintenance simpler by facilitating simple troubleshooting and enabling the replacement of defective components.
- **Safety:** Safety measures and shutdown procedures can be implemented using PLCs, improving workplace security.

PLCs are essential to modern industrial automation because they provide accurate and effective control over a variety of manufacturing operations and processes. They serve as the pillar of industrial control systems due to their adaptability, dependability, and capacity to carry out complex logic.

Human-machine interface (HMI)

An HMI is a user interface that links people to devices or processes, enabling users to communicate with and manage technology. The status, data, and controls of the underlying system are represented graphically or visually by HMIs, which makes it simpler for operators, engineers, and users to oversee, control, and communicate with complicated processes, machines, and equipment.

HMIs are essential in a number of sectors, including manufacturing, energy, transportation, and industrial automation. They improve operational management, data visualization, and decision-making by enabling effective communication between humans and machines.

The essential elements of HMIs

- **Graphical User Interface (GUI):** Information is presented visually in the GUI of an HMI using features like buttons, icons, graphs, charts, and color-coded indicators. It makes complex data and information easier to understand by simplifying it.
- **Touchscreen Display:** Many contemporary HMIs have touchscreen displays that let users tap, swipe, and drag items to interact with the interface.
- **Real-Time Data Visualization:** HMIs give users access to real-time data from sensors, gadgets, and systems so they can keep track of processes and events as they happen.

- **Control Elements:** With the use of buttons, switches, sliders, and other interactive features found on HMIs, users can issue orders to the linked system and manage its operations.
- **Alarms and Notifications:** HMIs provide faster operator response to urgent situations by displaying alerts, alarms, and notifications depending on specified conditions.
- **Trends and Historical Data:** Some HMIs allow users to analyze trends and performance over time by analyzing historical data.
- **Data Logging and Reporting:** HMIs can record and retain data for analysis and reporting, assisting with decision-making and performance evaluation.
- **Navigation and Menu Systems:** HMIs frequently have navigation menus to access various pages or areas of the interface, structuring the information.
- **Multilingual Support:** HMIs frequently handle numerous languages in global industries to accommodate users from various locales.

Types of HMIs

- **Traditional HMIs:** These are independent interfaces that show the state and controls of the system physically or virtually. They can be found in manufacturing facilities, industrial control rooms, and process control environments.
- **Touchscreen HMIs:** These HMIs are frequently utilized in applications spanning from industrial automation to consumer electronics and have interactive touchscreen displays.
- **Mobile HMIs:** HMIs can now be accessed by smartphones and tablets, enabling remote monitoring and control, thanks to the growth of mobile devices.

Benefits of HMIs

- **Operational Visibility:** HMIs give complex operations a visual representation, which improves operators' comprehension of system status and performance.
- **Efficient Control:** Through user-friendly interfaces, HMIs enable users to swiftly and precisely operate equipment and processes.
- **Real-Time Monitoring:** Users may keep an eye on events, alarms, and processes in real time, enabling quick decision-making.
- **Data-Driven Insights:** HMIs give users the ability to examine past data and data patterns to find chances for optimization.
- **Reduced Human Error:** By guiding users through activities and procedures, clear and intuitive interfaces help eliminate errors.

- **Remote Monitoring and Control:** Many HMIs allow for remote access, allowing users to keep an eye on and manage systems from any location.
- **Improved Safety:** HMIs can display alerts and safety warnings, assisting operators in successfully responding to urgent circumstances.

HMIs help people interact with and control sophisticated technology by bridging the gap between complicated systems and human operators. In many businesses that rely on automation and data-driven processes, their contribution to boosting operational efficiency, safety, and decision-making is crucial.

SCADA systems

Industrial control systems called Supervisory Control and Data Acquisition (SCADA) systems are used to monitor, manage, and control complex processes and operations across a variety of industries. SCADA systems enable operators to make educated decisions and optimize operations by providing real-time data visualization, remote monitoring, and centralized control over distributed processes.

Industries where effective process control and management are crucial include manufacturing, energy, water treatment, transportation, and more. SCADA systems are frequently employed in these sectors.

Key components and features of SCADA systems

- **Remote Terminal Units (RTUs) and Programmable Logic Controllers (PLCs):** RTUs and PLCs collect data from field sensors and equipment and send it to the SCADA system. Additionally, they get directives from the SCADA system to take actions in the field.
- **Human-Machine Interface (HMI):** The HMI is the user interface that engineers and operators utilize to view real-time data, alarms, and control options. Users can communicate with the SCADA system through it, keep track of operations, and take action.
- **Communication Protocols:** The central SCADA server, RTUs/PLCs, and field devices all communicate data using different communication protocols in SCADA systems. Typical protocols are DNP3, OPC, and Modbus.
- **Data Logging and Storage:** SCADA systems gather and preserve past data, enabling users to evaluate trends, assess performance, and make wise decisions.
- **Alarm Management:** Real-time alarm warnings and alerts are provided by SCADA systems based on specified thresholds or conditions, allowing operators to react to urgent situations quickly.

- **Trends and Graphical Representation:** SCADA systems frequently have tools for displaying historical visualizations, graphs, charts, and data trends to aid in analysis.
- **Historian Database:** SCADA systems frequently use historian databases to store plenty of old data for long-term reporting and analysis.
- **Security:** Security measures are implemented by SCADA systems to guard against illegal access, data breaches, and online threats.
- **Remote Monitoring and Control:** SCADA systems reduce the requirement for on-site presence by enabling operators to monitor and control processes remotely.

Benefits of SCADA systems

- Centralized Control
- Real-Time Monitoring
- Data-Driven Decision-Making
- Remote Access
- Reduced Maintenance Costs
- Regulatory Compliance

As SCADA systems have developed, they now include cutting-edge technology like cloud computing, edge computing, and advanced analytics. These developments strengthen their skills and establish them as essential components of sectors aiming for effective, data-driven operations.

Industrial robots

Industrial robots are adaptable tools made to automate a variety of tasks in a variety of industries, from production and assembly to material handling and precise operations. These robots are designed to carry out repeated, exact, and frequently complex operations with a high degree of accuracy, consistency, and efficiency. Modern production and automation are greatly aided by industrial robots, which raise productivity and quality.

Key characteristics and components of industrial robots

- **Manipulator Arm:** An industrial robot's manipulator arm is its main mechanical part. It has numerous links or segments joined by joints, allowing the robot to move in various directions.
- **End Effector (Gripper or Tool):** The robot's end effector is the component that interacts with the environment or the workpiece. To complete particular jobs, grippers, tools, or sensors can be fastened to the end effector.

- **Joints:** Robots frequently feature several joints that provide them the ability to move in a variety of directions, including rotating, translating, and bending. Revolute (rotational) and prismatic (linear) joints are common types of joints.
- **Controllers:** Specialized controllers are used to operate industrial robots. They carry out pre-programmed commands, govern the robot's movements, and oversee various elements of its operation.
- **Sensors:** For feedback and to interact with the environment, robots can be fitted with a variety of sensors, including cameras, force sensors, and proximity sensors.
- **Programming Interface:** Programming languages such as Robot Programming Language (RPL) and G-code, teach pendants, graphical user interfaces, and other techniques can be used to create robots.

Types of industrial robots

- **Articulated Robots:** These robots resemble a human arm in that they contain several rotational joints. They are quite adaptable and versatile and are frequently used for welding, painting, and assembling.
- **SCARA Robots:** SCARA is an abbreviation for "Selective Compliance Assembly Robot Arm." SCARA robots are renowned for their precise and quick movements and have horizontal joints. They are frequently utilized in pick-and-place and assembly applications.
- **Cartesian Robots:** Cartesian robots, also referred to as gantry robots, move along three linear axes (X, Y, and Z). Applications requiring both high precision and heavy lifting are ideal for them.
- **Delta Robots:** Delta robots are made for quick pick-and-place operations. They are utilized in sectors including packaging and food processing and have a special parallel-link structure.
- **Collaborative Robots (Cobots):** Cobots are created to collaborate with others in a common workspace. They are frequently employed for jobs requiring human-robot collaboration, like assembly and material handling, and have built-in safety precautions.

Applications of industrial robots

- Manufacturing and Assembly
- Material Handling
- Welding and Fabrication
- Packaging and Palletizing
- Machine Tending
- Automated Testing

Benefits of industrial robots

- Increased Productivity
- Consistency and Accuracy
- Labor Savings and Safety
- 24/7 Operation and Flexibility
- Quality Improvement

Motion control systems

Systems for controlling and directing the motion of machines, mechanisms, robots, and other mechanical systems are known as motion control systems. These systems guarantee exact control over the movement of moving parts' speed, position, and acceleration, enabling accurate and effective operation across a range of industries, including manufacturing, robotics, automotive, aerospace, and more.

To achieve desired motion profiles and performance, motion control systems use a combination of hardware elements, software algorithms, and feedback mechanisms. They are essential for providing repeatable and accurate movement in intricate applications (Figure 1.4).

Types of motion control systems

- **Open-Loop Motion Control:** Open-loop systems lack feedback and instead base the control action purely on the input instruction. They are utilized when great precision is not necessary or when external circumstances have little impact on the operation.
- **Closed-Loop Motion Control:** Sensor feedback is used in closed-loop systems to modify the control action and produce precise motion. They are frequently employed in situations when accuracy is essential.
- **Servo Motion Control:** To achieve precise positioning and speed control, servo systems use servo motors and feedback mechanisms.

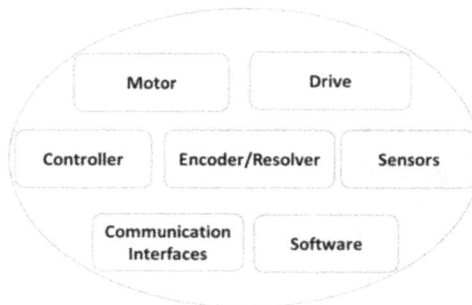

Figure 1.4 Key components of motion control systems.

They are utilized in high-precision applications including robotics and CNC machines.

- **Stepper Motion Control:** Stepper motors that move in distinct steps are used in stepper systems. They are utilized in applications like 3D printers and automated stages where precise placement is necessary.
- **Linear Motion Control:** In applications like material handling and assembly lines, linear motion control systems manage the movement of linear actuators or stages.
- **Robotic Motion Control:** In robotic systems, motion control is crucial to regulating the motion of numerous joints and achieving coordinated motion.

Benefits of motion control systems

- Precision and Versatility
- Repeatability
- Speed Control
- Energy Efficiency
- Reduced Wear and Tear
- Increased Throughput

Automated guided vehicles

Mobile robotic vehicles known as AGVs are used in industrial and commercial settings to move and carry materials, goods, or products without the assistance of humans. AGVs may carry out duties including material handling, transportation, and even some simple manipulation by navigating predetermined courses, avoiding obstacles, and using a variety of navigation systems and sensors. AGVs are frequently utilized to automate material transportation, increase efficiency, and lower labor costs in sectors like manufacturing, warehousing, logistics, and distribution centers (Figure 1.5).

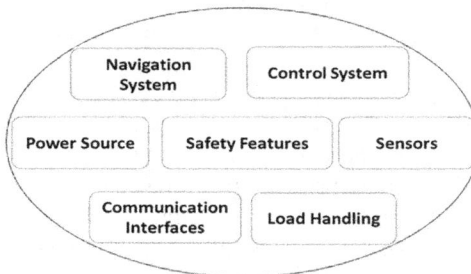

Figure 1.5 Key components of AGVs.

Types of AGVs

- **Tow Vehicles:** Trailers or carts loaded with materials can be moved using tow AGVs. They are frequently employed in manufacturing and logistical settings.
- **Unit Load AGVs:** Within warehouses and distribution facilities, these AGVs are made to transport pallets, containers, and other unit loads.
- **Forklift AGVs:** Pallets or other items can be lifted and transported by forklift AGVs, which perform the same tasks as manned forklifts.
- **Pallet AGVs:** AGVs with a focus on pallets are used to move and stack pallets in warehouses and distribution facilities.
- **Automated Storage and Retrieval System (AS/RS) AGVs:** These AGVs transfer commodities to and from storage locations in conjunction with automated storage systems.
- **Guided Vehicles with Manipulators:** Some AGVs have robotic arms or manipulators that allow them to carry out operations including picking, placing, and assembly.

Benefits of AGVs

- Labor Savings
- Efficiency
- Accuracy
- Safety
- Flexibility
- Space Optimization
- Data Collection

HOME AUTOMATION

Home automation, commonly referred to as smart home technology, is the process of connecting and automating numerous systems, appliances, and devices within a home [12–17]. Home automation gives homeowners remote control and automated procedures to monitor and operate their homes, improving convenience, energy efficiency, security, and comfort (Figure 1.6).

Examples of home automation

- **Smart Lighting:** Remote control and automatic on/off scheduling are both options for lighting. On the basis of user preferences or the amount of natural light, certain systems can change the light intensity.
- **Smart Thermostats:** To save energy, these gadgets adapt the temperature based on user preferences. They are frequently integrated with weather forecasts and can be operated remotely.

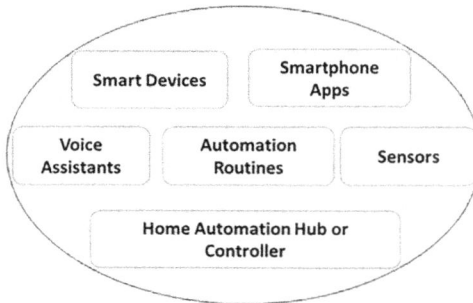

Figure 1.6 Key components of home automation.

- **Security and Surveillance:** Real-time monitoring and alarms are pro-vided through smart cameras, doorbell cameras, and motion sen-sors. On their smartphones, users may monitor camera feeds and get notifications.
- **Smart Locks:** Users who have remote-controllable locks at their doors can lock or open them from anywhere. They may also provide visitors brief access.
- **Smart Appliances:** It is possible to remotely control and see appliances like refrigerators, stoves, and washing machines. Some appliances even have alert systems for when maintenance is needed.
- **Energy Management:** Depending on occupancy or the time of day, home automation systems can optimize energy use by regulating light-ing, heating, and cooling.
- **Entertainment Systems:** For centralized control, smart TVs, audio systems, and streaming gadgets can be incorporated into automation systems.

Benefits of home automation

- **Convenience:** Remote access to home equipment allows users to man-age their living area more easily.
- **Energy Efficiency:** Automation can reduce utility costs and environ-mental effect by optimizing energy use.
- **Security:** Real-time monitoring and alarms are provided by smart security systems, which improve home security.
- **Comfort:** Automation can produce customized comfort settings, such as modifying the lighting and temperature to the user's preferences.
- **Peace of Mind:** Remote access and monitoring give homeowners the peace of mind they need by enabling them to watch over their house while they're away.

Smart devices

Everyday objects or appliances that have been improved with digital connectivity, intelligence, and the capacity to communicate and interact with other devices, users, or systems are referred to as smart devices, sometimes known as smart gadgets or smart appliances. With the help of technology like the IoT, sensors, data analytics, and wireless connectivity, these devices can make consumers' life more functional, convenient, and automated.

The notion of the smart home and the IoT as a whole, in which multiple devices are connected to form a space that responds to user preferences and commands, is the fundamental component of smart devices. They span a wide range of applications and range in size from tiny devices to substantial machines.

Examples of smart devices

- **Smartphones:** One of the most popular and functional smart devices, smartphones offer a variety of features beyond simple communication.
- **Smart Speakers:** Users can communicate with virtual assistants like Alexa, Google Assistant, or Siri to manage other smart devices, play music, and acquire information via gadgets like the Amazon Echo, Google Nest, and Apple HomePod.
- **Smart Locks:** These locks enable users to lock and open doors using cellphones or voice commands, providing remote control and monitoring of doors.
- **Smart Cameras:** With capabilities like motion detection and live video streaming, cameras like those from Ring and Arlo enable remote surveillance and monitoring.
- **Smart Appliances:** Smart refrigerators, washing machines, and ovens can all be operated and monitored from a distance, which is convenient and can save energy.
- **Fitness Trackers and Wearables:** Fitbit and Apple Watch, for example, track health and fitness parameters, check heart rate, and send notifications.

Benefits of smart devices

- Convenience
- Energy Efficiency
- Automation
- Remote Monitoring
- Data Insights
- Enhanced Functionality
- Customization
- Integration

Home automation hub or controller

A smart home system's central controller, also known as a hub, allows for the management and control of a variety of smart appliances and gadgets. It serves as a hub for communication, enabling users to access and manage all of their connected devices through a single user interface. By fusing several smart devices into one cohesive ecosystem, home automation hubs offer a simplified and seamless user experience.

Key functions and features of home automation hubs

- **Device Integration:** Users may operate lights, thermostats, cameras, door locks, and other smart home appliances from a single platform thanks to home automation hubs that facilitate the integration of multiple smart devices from different manufacturers.
- **Centralized Control:** With the help of hubs, users may manage numerous devices, program automated processes, and get notifications from a single interface.
- **Automation:** With the use of home automation hubs, users can design intricate automation scenarios in which certain events, timetables, or other circumstances cause certain actions to be taken.
- **Remote Access:** Using a smartphone app or a web-based interface, users may remotely control and monitor their smart devices, offering ease and flexibility.
- **Voice Control:** Numerous speech assistants, including Amazon Alexa, Google Assistant, and Apple Siri, are integrated with home automation hubs, allowing users to operate devices by speaking commands.
- **Interoperability:** By giving users a single platform for communication, hubs seek to close the communication gap between devices using various communication protocols.
- **Security:** To improve home security, hubs provide capabilities including remote monitoring, notifications, and warnings.
- **Customization:** Users can build scenes, customize automation, and change parameters to suit their preferences.
- **Energy Efficiency:** By coordinating the operation of devices like thermostats, lights, and appliances, hubs can assist in reducing energy use.

Types of home automation hubs

- **Cloud-Based Hubs:** These hubs link to a cloud server, making it possible to access and manage devices remotely from any location with an internet connection. Examples include Google Nest Hub and Amazon Echo Plus.

- **Local Hubs:** Local hubs run on the local network and don't rely much on the cloud. They provide more privacy and are not dependent on other servers.
- **DIY Hubs:** DIYers can build their own home automation hubs utilizing free and open-source software such as Home Assistant or OpenHAB.

Smartphone apps

Apps for smartphones and other mobile devices, such as tablets and smartphones, are also referred to as mobile apps. These apps provide a wide range of features and services to meet the demands of different user groups, including communication, productivity, entertainment, and utility. Smartphone applications offer customers a practical way to access and engage with digital services and content. They are made available through app stores and can be downloaded and installed onto a mobile device.

Types of smartphone apps

- **Social Networking Apps:** Users can interact, share content, and communicate with others via apps like Facebook, Instagram, and Twitter.
- **Messaging Apps:** Text, phone, and video communication are all available through messaging apps like WhatsApp, iMessage, and Telegram.
- **Utility Apps:** Tools for tasks like weather forecasting, taking notes, using a calculator, and scanning QR codes are included in utility apps.
- **Navigation and Maps Apps:** Real-time GPS navigation and traffic data are provided by navigation apps like Google Maps and Waze.
- **Entertainment Apps:** Gaming apps, streaming services (such as Netflix and Spotify), and multimedia consumption are all examples of entertainment applications.
- **Productivity Apps:** Users may manage projects, calendars, emails, and documents with the aid of productivity programs (such as Microsoft Office Suite and Evernote).
- **Health and Fitness Apps:** Applications like Fitbit, MyFitnessPal, and Apple Health provide wellness information, exercise regimens, and health tracking.
- **E-commerce and Shopping Apps:** Users of shopping apps can browse, shop, and buy products online (on eBay and Amazon).

Benefits of smartphone apps

- Convenience and Personalization
- Time Efficiency and Communication
- Entertainment and Remote Access
- Information Access

Voice assistants

Voice assistants, sometimes referred to as virtual assistants or AI assistants, are digital systems that recognize spoken commands and respond with useful information or carry out activities using natural language processing (NLP) and artificial intelligence (AI). Voice assistants make it easier and more comfortable for users to get information and carry out various tasks by allowing them to communicate with devices, services, and apps verbally.

Smartphones, smart speakers, smart TVs, and other connected gadgets frequently have voice assistants, enabling hands-free and voice-driven interaction for consumers.

Common voice assistants

- **Amazon Alexa:** Alexa was created by Amazon, powers Amazon Echo devices, and provides a huge selection of skills and integrations.
- **Google Assistant:** Google Assistant, developed by Google, is accessible on Android devices and built into Google Home smart speakers.
- **Apple Siri:** Available on iPhones, iPads, Macs, and other Apple devices is Apple's voice assistant.
- **Microsoft Cortana:** The speech assistant from Microsoft, Cortana, is integrated into Microsoft 365 and is available on Windows-based devices.
- **Samsung Bixby:** Samsung's voice assistant, Bixby, is available on Samsung smartphones and other electronics.

Benefits of voice assistants

- Hands-Free Interaction
- Convenience
- Accessibility
- Multitasking
- Automation

Automation routines

Automation routines are predetermined sets of operations that are pre-programmed to run automatically in response to certain conditions, events, or triggers. These procedures are intended to improve convenience, expedite processes, and lessen the need for manual involvement. Automation procedures are frequently utilized in a variety of contexts, such as digital workflows, commercial operations, and home automation.

Automation routines are used in the context of home automation to orchestrate and coordinate the behavior of numerous smart devices, resulting in a seamless and effective user experience. These routines can be programmed into smart home platforms or apps and are frequently activated by events like the time of day, changes in device status, or user interactions.

Examples of automation routines in home automation

- **Good Morning Routine:** The lights gradually come on, the thermostat is set to a pleasant setting, and the coffee maker begins brewing when the user's alarm goes off in the morning.
- **Movie Night Routine:** The user can start a "movie night" routine with a voice command via an app, which turns on the TV and sound system, dims the lights, and draws the window treatments.
- **Leaving Home Routine:** The smart thermostat changes to an energy-saving mode, all lights dim, and the security system turns on when the user leaves the house.
- **Bedtime Routine:** The lights go out, the doors lock, and the thermostat is turned to a cool setting when a user says "goodnight" to a voice assistant.
- **Security Routine:** The lights flash and a notification is sent to the user's smartphone when a motion sensor detects movement outside the home at night.

Business and digital workflows

To streamline procedures, decrease human error, and boost productivity, automation routines are also frequently employed in commercial and digital workflows. On the basis of predetermined criteria and conditions, these routines may comprise data processing, notifications, and task allocations.

Benefits of automation routines

- Time Savings
- Consistent and Efficient
- Reduced Errors
- Convenience
- Seamless Integration
- Remote Control
- Focus on Value-Added Tasks

Considerations to create automation routines

- **Clearly Define Objectives:** Establish the objectives and results you hope to achieve with automation.
- **Identify Triggers and Conditions:** Indicate the occasions or circumstances that will cause the routine to run.
- **Select Actions:** Choose the tasks or actions that should be carried out when the routine is triggered.
- **Test and Iterate:** Test the procedure to make sure it functions as expected and make changes as necessary.

- **Privacy and Security:** Make sure that automated processes and actions don't jeopardize security or privacy.
- **Review Regularly:** Review and update automated processes frequently to account for adjustments and changing requirements.

PROCESS AUTOMATION

By automating manual, repetitive, and rule-based procedures, process automation entails leveraging technology and software to improve and optimize corporate processes. Process automation aims to boost various company operations' efficiency, accuracy, consistency, and speed [18]. It crosses numerous sectors and industries, including manufacturing, customer service, human resources, and more.

Key concepts in process automation

- **Workflow Automation:** To do this, automated workflows that specify the order of actions, judgments, and interactions inside a business process must be designed and put in place. Tasks are completed in a predetermined order thanks to workflow automation, and data is easily transferred between phases of the process.
- **Robotic Process Automation (RPA):** RPA involves automating tasks within software applications utilizing software robots or bots [19]. RPA bots can do activities like data entry, data extraction, and data validation by imitating human interactions with user interfaces.
- **Business Process Management (BPM):** The analysis, design, optimization, and administration of business processes are all included in BPM. Modeling, automating, and monitoring processes using BPM software make them more effective and flexible.
- **Integration:** To facilitate seamless data exchange and collaboration across various stages of the business process, process automation sometimes entails connecting many systems and applications.
- **Data and Analytics:** A lot of data is produced through automation. Understanding the performance of the processes, identifying bottlenecks, and recognizing potential improvement areas through analysis are essential.

Examples of process automation

- **Invoice Processing:** Automating the procedure for processing invoices, from receipt of invoices to data validation, purchase order matching, and payment approval.
- **Employee Onboarding:** Automating the documentation, approvals, and system access provisioning steps in the onboarding process for new hires.

- **Customer Support:** Utilizing chatbots or automated systems to respond to typical customer questions and support requests.
- **Data Entry and Data Migration:** Automating data entry from one system to another, hence minimizing mistakes and manual labor.
- **Order Processing:** Automating all aspects of order fulfillment, including order entry, inventory control, and shipping.
- **Expense Reimbursement:** Automating the submission, evaluation, and authorization of employee expenditure reports.

Benefits of process automation

- **Increased Efficiency:** Automation speeds up throughput and lowers cycle times by reducing the amount of time needed to accomplish tasks and full processes.
- **Error Reduction:** Compared to human operations, automated systems are less prone to errors, improving data quality and accuracy.
- **Consistency:** Automated procedures make certain that work is carried out consistently in accordance with established guidelines and standards.
- **Resource Optimization:** Automating monotonous work frees up human resources, allowing them to concentrate on more strategic and value-added duties.
- **Compliance and Audit Trail:** Automated procedures can be created to guarantee adherence to rules and produce audit trails for accountability.
- **Scalability:** Increased workloads can be handled by automated processes without materially raising labor expenses.
- **Visibility and Insights:** Automation produces data that may be examined to learn more about the performance of the process and spot potential improvements.
- **Improved Customer Experience:** The customer experience can be improved by automation, which can result in quicker response times and more seamless interactions.

WORKFLOW AUTOMATION

By automating time-consuming, repetitive, and manual actions, workflow automation refers to the use of technology and digital tools to streamline and optimize corporate processes, tasks, and activities. By ensuring that tasks and data flow easily across different phases of a process, this automation seeks to increase efficiency, decrease errors, and foster collaboration.

Workflow automation entails the creation, execution, and administration of automated action sequences that adhere to predetermined logic and rules. It can be used in a variety of sectors and divisions, including marketing, customer service, human resources, and finance.

Examples of workflow automation

- **Employee Onboarding:** By automatically sending welcome emails, allocating responsibilities, creating contracts, and updating HR records, automation may speed up the onboarding process.
- **Invoice Processing:** Automated workflows can extract data from scanned documents, route invoices for approval, and update accounting software.
- **Lead Management:** Based on established criteria, automation technologies can qualify leads, send follow-up emails, and allocate leads to sales representatives.
- **Customer Support:** Automation can escalate serious issues, provide automatic solutions to common questions, and route client requests to the proper employees.
- **E-commerce Order Fulfillment:** Automated processes can create shipping labels, handle orders, adjust inventory levels, and email clients tracking information.

Benefits of workflow automation

- **Efficiency:** Automation decreases manual labor, accelerates workflow, and removes bottlenecks.
- **Accuracy:** Automation lowers the possibility of human error and inconsistent data.
- **Consistency:** Regardless of personal preferences, automated workflows ensure that procedures are followed consistently.
- **Productivity:** Instead of performing tedious administrative labor, staff may concentrate on tasks that bring value.
- **Data Visibility:** Real-time visibility into the condition and development of processes is made possible via automation.
- **Improved Collaboration:** By ensuring that information is shared seamlessly across team members, automation promotes improved cooperation.

ROBOTIC PROCESS AUTOMATION

RPA is a technique that uses software "bots" to automate routine, rule-based operations within corporate processes. By automating mundane processes that were previously completed by people, RPA enables firms to free up valuable human resources for more strategic and innovative work.

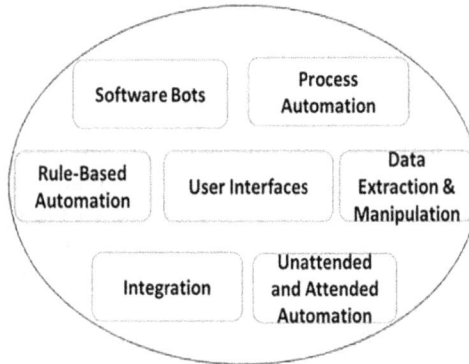

Figure 1.7 Key components of robotic process automation.

Similar to how humans interact with digital systems and applications, RPA bots carry out jobs reliably and quickly according to specified rules and procedures.

Processes including data entry, data extraction, data manipulation, and communication across various software systems are particularly well suited for RPA. Finance, customer service, human resources, and supply chain management are just a few of the departments and industries it can be employed (Figure 1.7).

Examples of RPA use cases

- **Data Entry and Validation:** RPA bots can transfer data accurately and with less manual effort from one system to another.
- **Invoice Processing:** Bots can validate data, update accounting systems, and extract data from invoices.
- **Customer Onboarding:** RPA can automate the procedures for gathering client data, checking papers, and opening accounts.
- **Claims Processing:** Insurance claims can be processed by bots by extracting data from claim forms, checking policy information, and updating claims systems.
- **Report Generation:** RPA bots can gather information from a variety of sources, produce reports, and send them to the appropriate stakeholders.

Business process management

BPM is a comprehensive strategy for overseeing and improving an organization's operational procedures to increase productivity, effectiveness, and adaptability. To align processes with business objectives, improve customer satisfaction, and boost overall performance, BPM entails analysis, design, implementation, monitoring, and continuous process improvement.

In an organization, BPM seeks to offer a structured framework for comprehending, recording, and optimizing both operational and strategic processes. To enable smooth process execution and ongoing improvement, it entails collaboration between numerous departments, stakeholders, and IT systems.

Key concepts of BPM

- **Process Analysis**: BPM starts by analyzing current processes to find bottlenecks, inefficiencies, and areas for development. To do this, a process's participants, tasks, and interactions must be mapped out.
- **Process Design**: Processes are changed or re-engineered in accordance with the study to remove bottlenecks, reduce procedures, and incorporate best practices. Roles, responsibilities, and decision points are all defined as part of the process design.
- **Process Modeling**: To depict processes and their interactions, BPM frequently uses visual models such as flowcharts, process diagrams, and swimlane diagrams.
- **Process Automation**: Using technologies like workflow systems, business rules engines, and RPA, tasks and activities are automated as part of BPM.
- **Workflow Management**: The orchestration of tasks, activities, and participants inside a process is a component of BPM. Task assignments, notifications, approvals, and transitions between process phases are managed by workflow systems.
- **Integration**: BPM combines several software programs, databases, and systems to provide efficient data transfer and communication among various process components.
- **Process Monitoring**: BPM systems use dashboards, metrics, and key performance indicators (KPIs) to provide real-time visibility into the performance of processes.
- **Process Optimization**: Constant improvement is a key component of BPM. Organizations regularly assess the performance of their processes and look for chances to improve and hone them for better results.

Benefits of BPM

- Efficiency and Consistency
- Improved Visibility
- Agility
- Customer Satisfaction

- Data-Driven Decision-Making
- Compliance
- Innovation and Cost Reduction

Integration

In the context of technology and business, integration is the process of integrating various systems, applications, or software parts to function as a single, well-coordinated unit. Integration makes it possible for many systems to seamlessly communicate data, information, and functionality, increasing productivity, accuracy, and overall business operations.

In today's interconnected and complicated business environment, where firms frequently use a range of specialized software tools to manage various elements of their operations, integration is essential. Eliminating data silos, cutting down on manual data entry, and improving information flow throughout the organization are all benefits of integrating these technologies.

Key concepts and approaches to integration

- **Application Programming Interface (API) Integration:** APIs offer a standardized method for data sharing and communication across various software applications. Through the use of API integration, systems can communicate by making requests and getting responses.
- **Middleware:** Middleware serves as a connecting layer that makes it easier for various applications and systems to communicate with one another. Features like data transformation, routing, and security are frequently offered.
- **Enterprise Service Bus (ESB):** A software architecture known as an ESB facilitates centralized communication and integration between diverse applications and services.
- **Data Integration:** Data integration is the process of merging information from several sources to create a complete and accurate picture. Techniques like data purification, transformation, and replication can be used in this.
- **Process Integration:** To develop end-to-end workflows, process integration focuses on coordinating and linking the business activities of various platforms.
- **Cloud Integration:** Cloud integration combines on-premises systems with cloud-based services and applications to ensure smooth data transfer between settings.

Benefits of integration

- Data Accuracy
- Time Savings

- Improved Decision-Making
- Enhanced Customer Experience
- Innovation and Scalability

Challenges and considerations

- **Data Security:** Making sure that sensitive data is safeguarded while being transmitted and stored is necessary for system integration.
- **Compatibility:** The usage of various technologies, data formats, and protocols by various systems might make integration attempts more difficult.
- **Maintenance:** To keep them operating and current, integrated systems need regular maintenance.
- **Vendor Support:** If there are third-party apps involved, integration may require cooperation and support from many vendors.

Data and analytics

In the current digital era, data and analytics are essential elements for many sectors of business, science, technology, and decision-making. The breakdown of these ideas is as follows:

- **Data:** Raw facts, statistics, or other types of information gathered through tests, measurements, surveys, or other methods are referred to as data. Various formats, including text, numbers, pictures, videos, and any other organized or unstructured format, are all possible. There are two main categories of data:
  - **Structured Data:** This kind of information is well-organized and neatly fits into pre-established categories or tables. It is frequently found in databases and is simple to process using conventional techniques.
  - **Unstructured Data:** Unstructured data does not cleanly fit into rows and columns and has a defined format or organization. Examples include written texts, messages on social media, pictures, audio files, and more. Advanced methods like computer vision and NLP are frequently needed when analyzing unstructured data.
- **Analytics:** Analytics entails the methodical analysis of data to find relevant insights, trends, patterns, correlations, and other significant information that can support decision-making. Several forms of analytics can be broadly grouped:
  - **Descriptive Analytics:** This kind of analytics concentrates on condensing historical data to comprehend what has previously occurred. It offers a look back at past occurrences and trends.

- **Diagnostic Analytics:** Diagnostic analytics looks for the causes of previous occurrences. It entails studying data to identify the factors that led to particular outcomes or patterns.
- **Predictive Analytics:** Making forecasts about upcoming occurrences or trends entails analyzing historical data and statistical algorithms, which is known as predictive analytics. Techniques from machine learning are frequently used in this kind of investigation.
- **Prescriptive Analytics:** This kind of analytics does more than just forecast the future. It offers suggestions about how to proceed to get the results you want. It involves recommending the optimal course of action in light of the available data.
- **Data and Analytics in Business:** In the context of business, data and analytics are used to better understand customer behavior, streamline processes, develop products and services, and spot untapped market niches. Businesses gather enormous volumes of data from a variety of sources, including social media, sales transactions, and consumer contacts, and use analytics to derive actionable insights from this data.

REFERENCES

1. Bangemann, T., Karnouskos, S., Camp, R., Carlsson, O., Riedl, M., McLeod, S., ... & Stluka, P. (2014). State of the art in industrial automation. *Industrial Cloud-Based Cyber-Physical Systems: The IMC-AESOP Approach*, 23–47.
2. Neumann, P. (2007). Communication in industrial automation – What is going on? *Control Engineering Practice*, *15*(11), 1332–1347.
3. Vyatkin, V. (2013). Software engineering in industrial automation: State-of-the-art review. *IEEE Transactions on Industrial Informatics*, *9*(3), 1234–1249.
4. Shell, R. (2000). *Handbook of Industrial Automation*. CRC Press.
5. Jammes, F., & Smit, H. (2005). Service-oriented paradigms in industrial automation. *IEEE Transactions on Industrial Informatics*, *1*(1), 62–70.
6. Gupta, A. K., & Arora, S. K. (2009). *Industrial Automation and Robotics*. Laxmi Publications.
7. Givehchi, O., Trsek, H., & Jasperneite, J. (2013, September). Cloud computing for industrial automation systems – A comprehensive overview. In *2013 IEEE 18th Conference on Emerging Technologies & Factory Automation (ETFA)* (pp. 1–4). IEEE.
8. Manesis, S., & Nikolakopoulos, G. (2018). *Introduction to Industrial Automation*. CRC Press.
9. Breivold, H. P., & Sandström, K. (2015, December). Internet of things for industrial automation – Challenges and technical solutions. In *2015 IEEE International Conference on Data Science and Data Intensive Systems* (pp. 532–539). IEEE.

10. Dey, C., & Sen, S. K. (Eds.). (2020). *Industrial Automation Technologies*. CRC Press.

11. Efe, M. Ö. (2011). Fractional order systems in industrial automation – A survey. *IEEE Transactions on Industrial Informatics*, 7(4), 582–591.

12. Brush, A. B., Lee, B., Mahajan, R., Agarwal, S., Saroiu, S., & Dixon, C. (2011, May). Home automation in the wild: Challenges and opportunities. In *Proceedings of the SIGCHI Conference on Human Factors in Computing Systems* (pp. 2115–2124).

13. Asadullah, M., & Raza, A. (2016, November). An overview of home automation systems. In *2016 2nd International Conference on Robotics and Artificial Intelligence (ICRAI)* (pp. 27–31). IEEE.

14. Gill, K., Yang, S. H., Yao, F., & Lu, X. (2009). A zigbee-based home automation system. *IEEE Transactions on Consumer Electronics*, 55(2), 422–430.

15. Alheraish, A. (2004). Design and implementation of home automation system. *IEEE Transactions on Consumer Electronics*, 50(4), 1087–1092.

16. Sriskanthan, N., Tan, F., & Karande, A. (2002). Bluetooth based home automation system. *Microprocessors and Microsystems*, 26(6), 281–289.

17. Al-Ali, A. R., & Al-Rousan, M. (2004). Java-based home automation system. *IEEE Transactions on Consumer Electronics*, 50(2), 498–504.

18. Love, J. (2007). *Process Automation Handbook: A Guide to Theory and Practice* (Vol. 42). London, UK: Springer.

19. Syed, R., Suriadi, S., Adams, M., Bandara, W., Leemans, S. J., Ouyang, C., ... & Reijers, H. A. (2020). Robotic process automation: Contemporary themes and challenges. *Computers in Industry*, 115, 103162.

20. Goldberg, K. (2011). What is automation? *IEEE Transactions on Automation Science and Engineering*, 9(1), 1–2.

21. Polo, M., Reales, P., Piattini, M., & Ebert, C. (2013). Test automation. *IEEE Software*, 30(1), 84–89.

22. Fewster, M., & Graham, D. (1999). *Software Test Automation* (pp. 211–219). Reading: Addison-Wesley.

23. Thummalapenta, S., Sinha, S., Singhania, N., & Chandra, S. (2012, June). Automating test automation. In *2012 34th International Conference on Software Engineering (ICSE)* (pp. 881–891). IEEE.

24. Wiklund, K., Eldh, S., Sundmark, D., & Lundqvist, K. (2017). Impediments for software test automation: A systematic literature review. *Software Testing, Verification and Reliability*, 27(8), e1639.

Chapter 2

RPA and test automation

Anil Kumar Dubey, Mala Saraswat,
and Erma Suryani

ROBOTIC PROCESS AUTOMATION

Software robots or bots are used in robotic process automation (RPA), a technology, to automate routine and rule-based processes within software applications. These bots are capable of imitating human behaviors on user interfaces, including button clicks, data entry, and mathematical operations. RPA is especially beneficial for activities that use structured data and adhere to predetermined routines.

Key concepts in RPA

- **Software Bots:** RPA bots are software programs that interact with already-existing software programs in a manner similar to that of a human user. They are able to move between screens, enter data, work with files, and retrieve information.
- **User Interface Interaction:** Without requiring changes to the underlying systems or databases, RPA bots interact with user interfaces in a manner similar to that of humans.
- **Rule-Based Automation:** RPA works best for jobs that have defined rules and stages. It is not intended for activities requiring deliberation or imaginative problem-solving.
- **Unattended and Attended Automation:** RPA can be attended or unattended, where the latter involves the bots collaborating with people and receiving input or direction as needed.
- **Workflow Automation:** RPA can integrate numerous jobs from various systems to automate whole workflows. For instance, it can retrieve information from emails, add it to a database, and provide reports.

Examples of RPA

- **Data Entry:** By copying data from one system or document and inputting it into another, RPA bots can automate data entry activities.

DOI: 10.1201/9781003479031-2

- **Invoice Processing:** Bots can check and input essential data that is extracted from invoices into accounting software.
- **Report Generation:** The process of compiling data from numerous sources, carrying out computations, and producing reports can all be automated by RPA.
- **Customer Onboarding:** During the onboarding process, bots can gather consumer information, do background checks, and input data into pertinent systems.
- **Order Processing:** By automatically entering orders, monitoring inventory levels, and updating order status, RPA helps speed up the processing of orders.

Benefits of robotic process automation

- **Efficiency:** RPA bots are always active, resulting in quicker task completion and higher throughput.
- **Cost Savings:** RPA lowers labor expenses and boosts cost-effectiveness by automating processes that were previously carried out by people.
- **Scalability:** Increased workloads can be handled by RPA without the need to add more staff members.
- **Compliance and Audit Trail:** RPA creates a thorough audit record of all actions and can be set up to adhere to regulatory and compliance regulations.
- **Faster Processing:** RPA may drastically shorten processing times, improving responsiveness and customer service.
- **Employee Satisfaction:** RPA frees up human employees to concentrate on more strategic and important work by taking over monotonous activities.

Considerations for RPA implementation

- **Process Analysis:** It's crucial to do a thorough analysis of the processes before using RPA to find jobs that can benefit from automation.
- **Security:** RPA bots must follow stringent security guidelines to guard against unwanted access to sensitive data.
- **Change Management:** It's crucial to use effective change management techniques to make sure that staff members are at ease with the adoption of RPA and comprehend its advantages.
- **Maintenance and Monitoring:** RPA bots must be continuously observed and maintained in order to maintain proper operation and react to environmental changes.
- **Integration:** RPA bots frequently need to communicate with different systems and programs, therefore appropriate integration is essential.
- **Vendor Selection:** Making the correct RPA software provider selection is essential for implementation success.

Software bots

Software bots, also known simply as "bots," are automated software programs created to carry out tasks that would typically be done by people. These bots are able to complete jobs repetitively and consistently, frequently far more quickly than people, and they can work continuously without the need for breaks. Customer service, data entry, information retrieval, process automation [1–9], and a host of other functions can all be developed into bots across a variety of sectors. Some essential characteristics of software bots are described.

Types of software bots

- **Chatbots**: These are computer programs created to converse with users, usually using text-based interfaces. Using natural language processing (NLP) methods, chatbots may answer queries, direct users through processes, and even imitate human-like conversations.
- **RPA Bots**: Routine and rule-based tasks inside corporate processes are automated using RPA bots. They can interface with software programs, work with data, and complete jobs that ordinarily call for human involvement.
- **Web Crawlers and Scrapers**: These bots are employed to harvest data automatically from websites. They access websites, gather information, and arrange it for analysis or other uses.
- **Social Media Bots**: On social media networks, some bots carry out operations such as posting content, liking, following, and interacting with users. However, some social media bots are employed in negative operations like spamming.
- **Gaming Bots**: Bots can replicate player behaviors in video games to automate gameplay or carry out activities to accomplish in-game goals.

User interface interaction

The use of visual and interactive components by users to interact with digital systems, applications, and software is referred to as user interface (UI) interaction. For a seamless and user-friendly experience that enables users to use the system to effectively accomplish their goals, a well-designed user interface is essential. An overview of UI interaction is shown below.

Elements of user interface interaction

- **Visual Elements**: These consist of elements such as menus, symbols, pictures, and typography. Users are guided through the interface by visual cues and information provided by visual elements.

- **Input Controls:** Users can submit data into the system using input controls. Text fields, checkboxes, radio buttons, drop-down menus, sliders, and date pickers are a few examples.
- **Navigation:** Users are assisted in navigating the system through navigational features. Menus, breadcrumbs, tabs, links, and search boxes are typical navigational components.
- **Feedback:** Users are informed about the results of their actions using feedback mechanisms. Success messages, problem messages, loading indications, and graphic adjustments in reaction to user inputs are all included in this.
- **Transitions and Animations:** By giving consumers visual indications regarding changes to the interface and assisting them in understanding the context of their activities, smooth transitions and animations can improve the user experience.
- **Microinteractions:** Microinteractions are brief, targeted interactions with a single goal. Toggling a setting, like a post, or confirming an action are some examples. These interactions help make the UI more engaging and usable overall.

UI interaction in different platforms

Web pages, mobile apps, desktop programs, and other digital interfaces can all benefit from UI interaction concepts. However, based on the capabilities of the device and user habits, each platform may have unique design principles and considerations. Emerging technologies such as voice interfaces, gesture controls, augmented reality (AR), and virtual reality (VR), which call for new design methodologies to ensure natural interactions, have also had an impact on UI interaction in recent years. In the end, UI interaction design that works improves user enjoyment, engagement, and effectiveness when utilizing digital goods and services.

Rule-based automation

Automation [10, 11] that is based on established rules or conditions is referred to as rule-based automation. With this method, particular activities are automatically carried out when certain criteria are satisfied. Numerous industries, including business, manufacturing, information technology, and others, employ rule-based automation extensively. It entails creating a logical framework so that, in the event that specific criteria are met, the appropriate actions can be taken automatically without human intervention.

Here's how rule-based automation works:

- **Rules Definition:** Defining the guidelines or requirements that must be met for automation is the first stage. Typically, "if-then" statements are used to describe these rules, allowing certain situations to be examined and associated actions to be established.

- **Condition Monitoring:** The system continuously checks to see if the predetermined conditions are met by monitoring the pertinent data or inputs. These characteristics may depend on elements such as time, events, data values, user activities, or any other pertinent aspects.
- **Triggering Actions:** The system initiates the corresponding actions when the circumstances outlined in the rules are met. Sending notifications, creating reports, updating databases, starting workflows, and other predetermined tasks are examples of these actions.
- **Automation Execution:** Based on the pre-established rules, the system performs the automated actions. Processes can be streamlined, efficiency can be increased, and the need for manual intervention can be greatly reduced.

Examples of rule-based automation

- **Email Filters:** Users of email applications frequently have the option to create rules that will automatically sort, categorize, or route emails depending on sender, subject, keywords, or attachments.
- **Home Automation:** To operate appliances such as lights, thermostats, and security cameras based on variables such as the time of day, occupancy, or sensor readings, smart home systems can use rule-based automation.
- **E-commerce:** Online retailers can set up rules that, when certain criteria, such as purchase amounts or consumer segments, are satisfied, discounts or promotions are automatically applied to particular products.
- **IT Operations:** To track network conditions, server performance, or security events, IT teams utilize rule-based automation. The system can provide alerts or initiate specified corrective actions when abnormal situations are found.
- **Manufacturing:** Rule-based automation can be used in industrial processes to regulate machinery and equipment depending on variables such as temperature, pressure, or measurements of the quality of the final product.

Unattended and attended automation

There are two different ways to deploy automation in business processes: unattended automation and attended automation. Based on the particular needs of the tasks and processes being automated, they are utilized for various objectives. Here is a description of both ideas:

Unattended automation

Automation of processes or operations that can be carried out without immediate human interaction is referred to as "unattended automation."

Usually repetitious and rule-based, these tasks involve little to no human decision-making. Back-end activities and operations that may be scheduled to execute at certain times or triggered by predefined criteria are frequently automated without human intervention. Unattended automation's salient features include:

- **No Human Involvement:** Unattended automation carries out activities automatically without requiring human intervention.
- **Scheduled Execution:** To avoid disturbances to regular business activities, automated tasks are scheduled to be executed at certain times, such as during off-hours.
- **Rule-Based:** Predefined rules or circumstances that specify how and when to carry out the tasks drive the automation.
- **High Volume:** Unattended automation is appropriate for jobs that require processing a lot of data or carrying out repetitive operations.

Examples: Unattended automation can be used to automate activities such as batch data processing, report production, data backups, server maintenance, and overnight data synchronization.

Attended automation

Attended automation, on the other hand, automates processes that need human intervention while still assisting and supporting human operators. The goal of attended automation is to increase the productivity of human workers by giving them the resources they need to carry out jobs more effectively. Collaboration between people and automation software is required. Attended automation's main aspects include:

- **Human Interaction:** Attendant automation combines automated tools with human interaction and decision-making.
- **Real-Time Assistance:** The human operator is assisted, guided, and supported by automation software as they carry out operations in real time.
- **User Initiated:** When necessary, the human operator activates the automation, and the software helps with task completion.
- **Enhanced Efficiency:** By utilizing automation while retaining human control, attended automation expedites work completion and lowers error rates.

Examples: Examples of attended automation include customer service representatives using chatbots to deliver prompt responses, data entry staff using automation technologies to validate and populate data fields, and financial professionals employing automated spreadsheet calculations.

Depending on the tasks involved, the extent of human engagement required, and the expected results, either attended automation or unattended automation should be chosen. While some jobs may benefit from attended automation to increase the productivity and accuracy of human workers, others may be totally automated utilizing unattended automation. To get the best outcomes, these strategies could occasionally be combined into a single process. The choice between attended and unattended automation depends on the particular context and objectives of the automation endeavor. Both types of automation improve business processes, lower errors, and boost productivity.

Workflow automation

Workflow automation is the process of streamlining and automating the steps, jobs, and procedures necessary to complete a particular commercial or operational workflow using technology. The objectives are to increase production, decrease manual intervention, reduce errors, and improve efficiency. Simple task automation and more elaborate end-to-end process automation can both be used in workflow automation. It can be used in a variety of fields and businesses, including manufacturing, healthcare, customer service, and finance.

Examples of workflow automation

- **Employee Onboarding:** Automating the onboarding of new employees by creating accounts, sending welcome letters, creating contracts, and assigning training modules.
- **Invoice Processing:** Automating the processing of bills through data extraction and validation, data routing for approval, and system updates for accounting.
- **Order Fulfillment:** Tracking inventory, updating stock levels, creating packing lists, and issuing shipment notifications to automate the order fulfillment process.
- **Customer Support:** Automating customer care procedures through the classification and routing of inbound support requests, the delivery of automated responses, and the referral of challenging issues to human agents.
- **Data Entry and Validation:** Automating data input duties through information extraction from documents and forms, database creation, and accuracy validation.
- **Marketing Campaigns:** Automating marketing efforts through audience segmentation, email scheduling, response tracking, and strategy modification based on user interactions.

IT AUTOMATION

IT automation is the process of streamlining and simplifying operations, procedures, and tasks that are related to IT using technology and software solutions. IT automation aims to increase productivity, decrease manual labor, increase accuracy, and enable quicker reaction to shifting business requirements. Information technology tasks and activities such as server management, network configuration, software deployment, monitoring, and other operations are all covered by IT automation.

Key aspects of IT automation

- **Configuration Management:** To maintain consistent configurations across servers, networks, and devices, automation solutions can be used, which lowers the possibility of errors and increases security.
- **Provisioning and Deployment:** By automating the provisioning of servers, virtual machines, and software applications, downtime is reduced and human error is eliminated while also ensuring speedy and uniform setup.
- **Orchestration:** Multiple automation tasks are brought together through orchestration into workflows that make sure procedures are carried out in the right order and in accordance with established criteria.
- **Monitoring and Alerting:** Real-time monitoring of IT applications and infrastructure can be facilitated by automation, resulting in warnings and solutions when problems are found.
- **Patch Management:** Application of software patches and upgrades across systems is automated to keep the environment current and safe.
- **Backup and Recovery:** IT automation can manage backup procedures, schedule them, and help with disaster recovery situations.
- **DevOps Automation:** Automation is essential for continuous integration and continuous deployment (CI/CD) pipelines in software development, enabling quicker and more dependable product releases.

Examples of IT automation

- **Server Provisioning:** Automating the process of installing and configuring the required operating systems and applications on servers, whether they are physical or virtual.
- **Network Configuration:** Automating network device configuration to ensure uniform and precise network settings, such as routers and switches.

- **Software Deployment:** Guaranteeing consistency and minimizing manual labor by automating the deployment of apps and updates across diverse systems and environments.
- **Monitoring and Alerts:** Establishing automated monitoring systems to track the performance and health of servers, apps, and network devices and to send out alerts when problems occur.
- **Security Management:** Automating security procedures such as access control, vulnerability assessments, and user access management.
- **Backup and Recovery:** Offering effective recovery procedures in the event of data loss or system failure and automating routine backups.

Benefits of IT automation

- **Reduced Errors:** Automating repetitious procedures and manual settings reduces human error.
- **Faster Response:** Rapid reactions to problems and incidents are made possible by automated monitoring and alerting systems.
- **Scalability:** Without a lot of human labor, automation enables IT systems to scale quickly to handle increased workloads.
- **Resource Optimization:** Instead of performing routine maintenance and configuration, IT staff can concentrate on more strategic and valuable duties.
- **Compliance:** Consistent compliance with rules and company policies can be enforced through automation.
- **Agility:** Automation facilitates more rapid adjustment to shifting corporate requirements and technology developments.

Configuration management

Software engineering and systems engineering discipline known as configuration management is concerned with managing and regulating changes to systems, software, hardware, and other components during the course of their lifecycles. Configuration management's fundamental objective is to ensure that products and systems are created, maintained, and delivered in a controlled and consistent manner while reducing risks and errors. Maintaining the integrity and dependability of complex systems, especially as they change over time, requires configuration management.

Key aspects of configuration management

- **Configuration Identification:** This entails defining and identifying the distinct parts of a system or product. Each component is given a special identification number, and the connections and dependencies between them are listed. This step creates the framework for successfully handling changes.

- **Configuration Control:** Changes to system components are managed and approved through a formal process thanks to configuration control. Examining proposed modifications, gauging their effects, and deciding whether to accept and execute them are all steps in this process.
- **Configuration Status Accounting:** The process of configuration status accounting entails keeping meticulous records of the status and development of each system component. This enables stakeholders to monitor changes, comprehend the current setup, and evaluate how changes will affect the system as a whole.
- **Configuration Verification and Audit:** Verification and audit procedures make sure that a system's or product's actual state corresponds to its intended configuration. Periodically, audits are carried out to ensure that the paperwork and components correspond to the stated criteria.
- **Change Management:** The methods for proposing, analyzing, approving, and implementing changes are governed by change management processes. These procedures aid in preventing unapproved or uncontrolled modifications that can cause mistakes or disruptions.

Provisioning and deployment

Two essential ideas in the field of IT and software development are provisioning and deployment. For systems to be functional and user-accessible, they entail setting up, configuring, and distributing resources, applications, and services. These ideas deal with several phases of the software development and operations lifecycle, despite the fact that they are related.

- **Provisioning:** The process of setting up and configuring the resources, including hardware, software, network resources, and other infrastructure components, is referred to as provisioning. This enables the operation of applications or services. It entails laying the groundwork for the deployment and operation of applications. Physical hardware, virtual computers, cloud resources, containers, and more can all be provisioned. Provisioning frequently involves the automatic production and allocation of resources based on predetermined criteria in the context of cloud computing. For instance, resources such as virtual computers, storage, and networking components can be deployed programmatically using APIs and templates in Infrastructure as a Service (IaaS).
- **Deployment:** The process of making the software or program created during the development phase accessible and functional for end users in a particular environment is known as deployment. This entails setting up the program on the selected infrastructure, be it on-premises servers, cloud servers, or other platforms, and running it as it is configured and installed. Assembling databases, establishing

Table 2.1 Provision vs. deployment

Key feature	Provisioning	Deployment
Focus	Focuses on setting up the necessary resources and infrastructure for applications to run	Focuses on taking the developed application and making it operational in a specific environment
Timing	Occurs before deployment and involves configuring the environment to support the application	Occurs after provisioning and involves installing and configuring the application itself
Scope	Encompasses the entire infrastructure, including hardware, software, and networking resources	Primarily focuses on the application code, databases, and related configurations
Automation	Often involves automated processes to create and configure resources based on defined specifications	Can also involve automation for tasks such as application installation, configuration, and scaling
Example	Creating virtual machines, setting up networking, and allocating storage resources in a cloud environment	Installing and configuring a web application on the provisioned virtual machines, ensuring proper database connections, and making the application accessible to users

application servers, integrating with external services, maintaining security precautions, and making the program user-accessible are all included in the deployment process. Deployment can be a challenging procedure, especially for multi-component, large-scale programs (Table 2.1).

Orchestration

In the context of IT and technology, orchestration refers to the coordination and automation of numerous tasks, processes, and components in order to accomplish a particular objective or result. Management and sequencing of several actions or services into a unified workflow is known as orchestration. To ensure effective and reliable completion of challenging tasks, this idea is widely employed in a variety of fields, including cloud computing, software development, networking, and business process management.

Examples of orchestration

- **Cloud Orchestration:** In the cloud, orchestration technologies automate and dynamically manage the provisioning, deployment, scaling, and monitoring of virtual machines, containers, and other resources.

- **DevOps Orchestration:** To provide software more quickly and reliably, DevOps approaches entail orchestrating several phases of software development, testing, deployment, and monitoring.
- **Business Process Orchestration:** Order processing, supply chain management, and customer onboarding are just a few examples of the end-to-end business processes that companies utilize orchestration to automate and streamline.
- **Network Orchestration:** In order to construct and sustain complex network architectures, network orchestration entails automating the configuration, provisioning, and administration of network resources.
- **Container Orchestration:** The deployment, scalability, and maintenance of containerized applications across clusters of hosts are orchestrated by tools like Kubernetes.

Orchestration vs. automation

Despite the fact that orchestration involves automating and coordinating actions, it usually works with more complicated circumstances involving numerous interconnected components. Contrarily, automation can refer to the automation of certain tasks or processes without the requirement for considerable inter-element coordination. Overall, by offering a mechanism to optimize and automate complicated processes and workflows, orchestration plays a critical role in managing the complexity of modern IT settings.

Monitoring and alerting

Critical elements of IT operations and systems management include monitoring and alerting. They support preserving the performance, availability, and dependability of networks, systems, and other infrastructure elements. While alerting entails contacting pertinent stakeholders when certain thresholds or circumstances are reached, monitoring requires continuously observing the health and state of these elements. These procedures are essential for spotting problems, avoiding downtime, and quickly responding to difficulties.

Monitoring: Continuous observation and data gathering about numerous elements of a system or application constitute monitoring. This information often relates to user activity, system health, resource use, performance indicators, and more. Monitoring enables IT teams to track key performance indicators (KPIs), spot trends, and obtain insights into the environment's current state. There are many different methods of monitoring, such as:

- **Performance Monitoring:** Monitoring system performance parameters such as CPU use, memory consumption, disk I/O, network traffic, and response times.

- **Availability Monitoring:** Ensuring that resources, services, and applications are accessible and responsive to user demands.
- **Application Monitoring:** Monitoring user interactions, behavior, and performance within certain applications to find problems and enhance user experience.
- **Security Monitoring:** Spotting and combating security threats, unauthorized access, and strange activity in the IT environment.

Alerting: When certain predetermined circumstances are met, alerting entails contacting pertinent persons or teams. These criteria may be thresholds for deviant conduct, declining performance, or prospective problems. Various communication methods, including email, text messaging, mobile apps, and dashboard notifications, are frequently used to send alerts. Alerts are intended to encourage timely action to prevent or reduce hazards. Monitoring and alerting are closely related because alerts are based on observed data.

Key concepts and benefits

- **Real-Time Visibility:** Real-time visibility into the functionality and state of systems and applications is provided via monitoring.
- **Early Detection:** Before they become significant problems or cause downtime, monitoring helps find faults and abnormalities.
- **Root Cause Analysis:** Monitoring data can assist in identifying the underlying causes of performance problems, enabling IT teams to more efficiently resolve issues.
- **Proactive Maintenance:** Alerts make IT staff aware of potential issues, enabling them to take preventative action to avoid disruptions.
- **Incident Response:** Critical events are immediately alerted by alerting, allowing for quick action and resolution.
- **Capacity Planning:** By recognizing resource utilization trends and forecasting future needs, monitoring aids in capacity planning.

Integration with incident management

Practices for incident management are closely related to monitoring and alerting. Incidents are unanticipated setbacks or problems that interfere with a system's or service's regular operation. By providing pertinent information and facilitating quick action, monitoring and alerting play a crucial part in identifying, responding to, and resolving issues. Monitoring and alerting technologies are crucial parts of a proactive strategy for managing and maintaining complex systems, ensuring peak performance, and providing users with dependable services in modern IT environments.

Patch management

Planning, testing, implementing, and monitoring changes, or patches, to software programs, operating systems, and other technological components is the process of patch management. Software providers release patches to remedy security flaws, correct problems, boost performance, and add new features. The security and stability of IT environments depend on effective patch management since outdated software leaves them vulnerable to threats like cyberattacks and data breaches.

Key components of patch management

- **Patch Identification:** The first stage entails determining whether software and system patches are accessible. Monitoring vendor websites, security alerts, and automatic patch management systems may be necessary.
- **Patch Assessment:** To identify the relevance and significance of patches to the environment, organizations evaluate the impact and urgency of the patches. Critical security patches are frequently given priority for deployment right away.
- **Testing:** Patches should be tested in controlled conditions before being used in production settings to make sure they don't have any compatibility problems or unforeseen effects.
- **Deployment:** Patches are deployed to the relevant systems after they have been examined and authorized. The deployment techniques used in this procedure might be manual or automated.
- **Monitoring and Verification:** Systems should be watched once patches are applied to make sure they are working properly and haven't caused any new problems.
- **Documentation:** For audit purposes, thorough patch management process documentation, including patch distribution dates, testing outcomes, and deployment logs, is crucial.

Patch management in different environments

Patch management applies to various technology environments, including:

- **Operating Systems:** Protecting against vulnerabilities requires routinely applying security patches to operating systems (including Windows, Linux, and macOS).
- **Applications:** To address security problems, programs such as web browsers, office suites, and other software require routine updating.
- **Servers:** Patching server systems on a regular basis is necessary to maintain the security and dependability of vital services.

- **Network Devices:** To maintain security and functioning, network infrastructure components such as routers, switches, and firewalls need to be patched.
- **Virtual Machines and Cloud Services:** Patch management is also necessary for cloud services and virtualized systems to prevent vulnerabilities.

Backup and recovery

Fundamental data management and information technology (IT) procedures include backup and recovery. In order to protect against data loss, system failures, disasters, and other unforeseen events, they include making copies of the data and the systems. In the event of data loss or a system breakdown, recovery entails restoring data from backups. Backup is the process of duplicating and storing data.

Backup: Backup entails producing copies of data, files, applications, and configurations for storage in a different location; these copies are frequently created on various physical devices or through cloud-based services. Backups are crucial for preventing data loss from hardware malfunctions, software bugs, unintentional deletions, cyberattacks, and natural catastrophes. There are various kinds of backups:

- **Full Backup:** A complete backup of all data and files, including any updates since the last backup, offers complete recovery possibilities, but may need additional time and storage.
- **Incremental Backup:** Only data that has changed or been added since the last backup is transferred, compared to full backups, requires less time and storage, but may involve more difficult recovery procedures.
- **Differential Backup:** It transfers all data that has changed since the last complete backup, unlike incremental backup. Easy to restore compared to incremental backups, but may eventually need more storage space.
- **Snapshot:** A snapshot copy of the data and systems. For rapid recuperation, snapshots can be taken at regular intervals.

Recovery: When there has been a loss of data, a system breakdown, or other occurrence, recovery entails restoring data from backup copies. Data loss prevention, system restoration, and downtime reduction are the main objectives. Recovery procedures may consist of:

- **Data Recovery:** Individual files, databases, or data sets can be restored from backup copies.

- **System Recovery:** Bringing back to life the complete operating system, applications, and configurations.
- **Disaster Recovery:** Data and system restoration following major events such as hardware malfunctions, fires, floods, or cyberattacks.

Key concepts and benefits

- **Redundancy:** Backups reduce the risk of data loss by maintaining several copies of the data.
- **Data Protection:** Backups shield information from numerous risks, including as malware, hardware malfunctions, and human error.
- **Business Continuity:** After disruptive events, effective recovery procedures assist in keeping businesses operating.
- **Compliance:** To ensure data availability and integrity, many industries and regulations mandate data backup and recovery procedures.
- **Peace of Mind:** Consistent backups and clear recovery strategies give assurance that crucial data can be restored.

DevOps automation

DevOps automation is the process of streamlining and accelerating the development, deployment, and operation of software applications using automated tools, procedures, and practices. By increasing communication and reducing silos between the development and IT operations teams, DevOps strives to increase the speed and reliability of software delivery. By lowering the need for manual intervention, decreasing errors, and facilitating effective and standardized workflows, automation is essential to attaining these objectives.

Key areas of DevOps automation

- **Continuous Integration (CI):** Code changes from many developers are automatically integrated into a shared repository as part of CI automation. To find problems early in the development process, automated builds, tests, and code reviews are carried out.
- **Continuous Delivery/Deployment (CD):** By automating the distribution of tested code updates to diverse environments, CD automation expands on CI. This guarantees that code is ready for production and that it can be deployed fast and effectively.
- **Infrastructure as Code (IaC):** IaC entails utilizing code to automatically provide and manage infrastructure resources. This promotes infrastructure that is version-controlled, ensures consistency, and minimizes human configuration.

- **Configuration Management:** In order to control configuration drift and guarantee that systems are configured correctly, automation technologies are used to maintain and enforce consistent configurations across various environments.
- **Automated Testing:** Unit tests, integration tests, and regression tests are examples of automated testing that is carried out automatically as part of the CI/CD pipeline to find flaws and guarantee product quality.
- **Release Orchestration:** By organizing the many processes, approvals, and activities involved in deploying a new version of software, automation solutions assist in orchestrating the release process.
- **Monitoring and Alerting:** Applications and infrastructure are monitored, alarms are triggered based on predetermined thresholds, and automated actions to events are started using automation.

Benefits of DevOps automation

- Speed and Reliability
- Consistency and Efficiency
- Scalability and Collaboration

TEST AUTOMATION

The practice of automating the execution of tests on software systems, applications, or products is known as test automation [12–16]. Test automation seeks to improve testing effectiveness while lowering human error and ensuring test correctness and repeatability. Testers can concentrate on more intricate and creative testing tasks by using test automation, which is especially helpful for repetitive and time-consuming tests.

Test scripts

To verify the functionality, performance, and quality of a software application, test script automation involves automating the execution of software testing scripts. Test scripts are detailed sets of guidelines that specify the particular test cases, inputs, anticipated results, and verification standards. The execution of automated test scripts by automation tools simulates user interactions, data inputs, and system responses. Automated test scripts are created using scripting or programming languages. This automation aids in testing process simplification, efficiency enhancement, and testing activity accuracy improvement.

Key steps in test script automation

- **Test Case Selection:** Determine which test cases can be automated. The finest returns on investment for automation are frequently produced by complex and repetitive test cases.
- **Test Script Creation:** Use scripting or programming languages such as Python, Java, or JavaScript to create test scripts. For the purpose of writing scripts, automation frameworks and libraries can offer reusable elements and structures.
- **Test Data Preparation:** Prepare test inputs and data that will be used by automated scripts to mimic user behavior and system behavior.
- **Automation Tool Selection:** Depending on the application's technology stack, testing requirements, and resource availability, select an appropriate automation tool or framework.
- **Test Script Execution:** Run the scripts for the automated tests on the application being tested. The programs mimic user inputs, interactions, and activities.
- **Verification and Reporting:** Test automation creates reports outlining test results, including pass/fail status and any found flaws, and compares actual results to anticipated results.

Benefits of test script automation

- Efficiency and Better Accuracy
- Reusability
- Parallel Execution
- Consistency

Test frameworks

Standardized structures, rules, and procedures known as test frameworks serve as the base for creating, planning, and carrying out software testing. Test frameworks provide an organized method for creating and maintaining test cases, promoting testing process uniformity, maintainability, and scalability. In order to facilitate successful and efficient software testing, these frameworks offer tools and libraries that make it easier to create, run, and report tests.

Some common types of test frameworks

- **Unit Testing Frameworks:** Unit testing frameworks are made to test distinct parts, operations, or procedures of a software program. Examples include pytest (Python), JUnit (Java), NUnit (.NET), and Jasmine (JavaScript).

- **Functional Testing Frameworks:** Functional testing frameworks concentrate on testing the features and functionality of the application as a whole. Examples include TestCafe for web apps, Appium for mobile applications, and Selenium for web applications.
- **Integration Testing Frameworks:** Frameworks for integration testing examine how various application modules, components, or services interact with one another. Examples include Test::Unit (Ruby), unittest (Python), and TestNG (Java).
- **End-to-End (E2E) Testing Frameworks:** E2E testing frameworks model user interactions with applications to make sure that each component functions as intended. Cypress, Protractor, and Nightwatch.js are other examples.
- **Behavior-Driven Development (BDD) Frameworks:** Tests are easier to understand for non-technical stakeholders because of BDD frameworks, which employ natural language syntax to express the intended behavior of software components. Examples include Behave (Python), SpecFlow (.NET), and Cucumber (which supports a number of languages).
- **Performance Testing Frameworks:** Frameworks for performance testing assess an application's responsiveness, scalability, and overall performance under various circumstances. Examples include Gatling, Locust, and Apache JMeter.
- **Mocking and Stubbing Frameworks:** When testing, these frameworks make it easier to isolate particular components by making mock objects or stubs. Examples include unittest.mock (Python), Moq (.NET), and Mockito (Java).

Automated test suites

An organized group of automated test cases that are controlled and run together is referred to as an automated test suite. Collectively, these test cases focus on particular functionalities, features, or parts of a software application. An automated test suite is intended to speed up testing, increase effectiveness, and guarantee complete testing coverage across all application-related areas.

Benefits of automated test suites

- **Efficiency:** Compared to manual testing, automated test suites provide faster and more frequent testing.
- **Accuracy:** Automated testing reduces the possibility of human errors and inconsistencies because they are reliable and repeatable.
- **Regression Testing:** Regression testing uses test suites to make sure that code changes don't result in new problems.

- **Parallel Execution:** Parallel execution of automated test suites can save time and resources.
- **Continuous Integration and Deployment:** Continuous Integration and Continuous Deployment (CI/CD) pipelines depend on automated test suites to validate code changes prior to deployment.
- **Reusable Test Cases:** A test suite's test cases can be applied to many scenarios, contexts, and releases.

Components of an automated test suite

- **Test Cases:** Certain functionality or scenarios in individual automated test cases.
- **Configuration Management:** Environmental setups, settings, and configurations required for test execution.
- **Test Data:** Test cases use inputs and data sets to simulate various scenarios.
- **Test Framework Integration:** Integration with the tool or framework for test automation of choice.
- **Reporting and Logging:** Mechanisms for producing reports on how tests were executed and logging information for analysis.
- **Test Suite Management Tools:** Tools or programs to administer and run the test suite.

Regression testing

Regression testing is a sort of software testing that is concerned with ensuring that recent updates or enhancements to a software application do not negatively impact the features that are currently in place. Regression testing's major objective is to make sure that newly added code does not corrupt previously tested features or bring new bugs into the software.

Key concepts of regression testing

- **Code Changes:** Regression testing is done whenever the codebase is modified, whether it is for bug patches, new features, improvements, or optimizations.
- **Scope:** Regression testing covers the modified areas specifically as well as any related functionalities that may also be indirectly impacted by the modifications.
- **Automated Testing:** Regression testing frequently uses automated test cases since they can test a lot of scenarios fast and consistently.
- **Reusability:** Regression testing can be done more quickly and with less work by reusing existing test cases, test scripts, and test suites.

- **Continuous Integration (CI) and Continuous Deployment (CD):** To ensure that code changes are tested before being deployed to production, regression testing is a crucial component of CI/CD workflows.
- **Version Control:** Version control systems assist in identifying what has been changed and the resulting effects by keeping track of changes.

Benefits of regression testing

- **Bug Detection:** Regression testing ensures that the product is stable by identifying new flaws brought on by recent code changes.
- **Preserve Quality:** Regression testing maintains the software's quality by making sure that current features continue to function as intended.
- **Confidence in Changes:** Gaining trust that code modifications are secure and do not adversely affect the application gives developers and stakeholders confidence.
- **Faster Issue Identification:** Early problem detection during development decreases the time and effort needed to fix problems.
- **Support for Agile Practices:** By confirming changes made during each iteration or sprint, regression testing promotes iterative development processes.

Continuous integration/continuous deployment (CI/CD)

The goals of CI and continuous deployment (CD), two software development approaches, are to speed up the development process, enhance the quality of the program, and make it possible to send code changes to production settings with reliability and speed. They are frequently utilized in tandem as a component of CI/CD, a comprehensive strategy for software delivery.

Continuous Integration (CI): Continuous Integration is a technique where developers frequently integrate their code changes into a shared repository, including version control systems (like Git), numerous times daily. Every integration starts an automatic build process that includes compiling, testing, and validating the code updates. The fundamental ideas of CI include:

- **Frequent Integration:** To reduce integration problems and conflicts, developers integrate their code frequently.
- **Automated Builds:** The code changes are automatically compiled and built using automated build tools.
- **Automated Testing:** To verify the code changes, automated tests, such as unit tests and integration tests, are run.

- **Early Detection:** Early in the development phase, CI aids in the detection of integration problems, regressions, and flaws.
- **Version Control:** Version control systems are used to keep track of code alterations, enabling collaboration and history monitoring.

Continuous Deployment (CD): The automated build, test, and integration process is furthered by the CI extension known as Continuous Deployment. Continuous deployment makes changes to the code that pass the automated tests available to users immediately by automatically deploying them to production or staging environments. The fundamental tenets of CD include:

- **Automated Deployment:** Test-passing code modifications are automatically rolled out to production or staging environments.
- **Reduced Manual Steps:** To decrease errors and delays, manual intervention in the deployment process is minimized.
- **Rapid Release Cycle:** Rapid release cycles made possible by continuous deployment make it possible for users to immediately access new features and upgrades.
- **User Feedback:** Quick deployment enables quicker customer feedback, allowing developers to refine their work in response to user feedback.
- **Rollback and Monitoring:** The stability of production systems is guaranteed by automated monitoring and faster rollback capacity in the event of problems.

Benefits of CI/CD

- Faster Time to Market
- Reduced Risk
- Improved Collaboration
- Consistent Quality
- Continuous Improvement
- Agile Practices

Challenges and considerations

- **Infrastructure as Code (IaC):** For CI/CD to be successful, infrastructure provisioning and administration must be automated.
- **Testing Strategy:** To detect errors and regressions as soon as possible, robust automated testing techniques are required.
- **Security Considerations:** For deployments to be safe, security procedures should be incorporated into the CI/CD pipeline.
- **Cultural Shift:** Development, testing, and operations teams frequently need to undergo a cultural transformation in order to use CI/CD.

- **Rollback Plan:** In the event that a deployment produces problems in the production environment, a clear rollback strategy is crucial.

Load and performance testing

In the software development lifecycle, load testing and performance testing are essential tasks that evaluate an application's responsiveness, stability, and scalability under varied degrees of user demand and system stress. These tests are designed to find any potential bottlenecks, performance degradation, or other problems that may affect how the application behaves when used by many people or under load-intensive circumstances.

Load Testing: In order to assess an application's performance and behavior, load testing includes putting it to a certain load, which is often represented by a simulated number of concurrent users or transactions. The objective is to evaluate the application's capacity to manage the anticipated load and to spot any performance problems that may develop under various stress levels.

Benefits of load testing

- Performance Validation
- Capacity Planning
- Identifying Bottlenecks
- Scalability Testing

Performance Testing: Performance testing assesses the overall performance of an application in terms of responsiveness, speed, and stability under various circumstances. It includes a variety of measures, including load testing, stress testing, and endurance testing, to give a thorough picture of the performance characteristics of the application.

Key concepts of performance testing

- **Response Time:** Evaluates how quickly an application responds to user actions.
- **Throughput:** Determines how many transactions the application can handle in a given amount of time.
- **Stress Testing:** Apply a load that is greater than the application's anticipated capacity to test how it responds in difficult circumstances.

Benefits of performance testing

- **User Experience:** Performance testing makes certain that users receive quick responses and fluid interactions.

- **Capacity Analysis:** To identify the application's maximum capacity and scalability, performance testing is useful.
- **Reliability:** Identifies potential problems with performance that can cause crashes, slowdowns, or unresponsiveness.
- **Infrastructure Optimization:** Performance testing aids infrastructure optimization for effective resource usage.

To make sure that applications fulfill user expectations for speed, dependability, and responsiveness, load testing and performance testing are essential. By running these tests, you can find potential performance bottlenecks and make sure the application can properly manage a range of user demands.

REFERENCES

1. Jämsä-Jounela, S. L. (2007). Future trends in process automation. *Annual Reviews in Control, 31*(2), 211–220.
2. Van der Aalst, W. M., Bichler, M., & Heinzl, A. (2018). Robotic process automation. *Business & Information Systems Engineering, 60,* 269–272.
3. Love, J. (2007). *Process Automation Handbook: A Guide to Theory and Practice* (Vol. 42). London, UK: Springer.
4. Hofmann, P., Samp, C., & Urbach, N. (2020). Robotic process automation. *Electronic Markets, 30*(1), 99–106.
5. Chakraborti, T., Isahagian, V., Khalaf, R., Khazaeni, Y., Muthusamy, V., Rizk, Y., & Unuvar, M. (2020). From robotic process automation to intelligent process automation: Emerging trends. In *Business Process Management: Blockchain and Robotic Process Automation Forum: BPM 2020 Blockchain and RPA Forum, Seville, Spain, September 13–18, 2020, Proceedings 18* (pp. 215–228). Springer International Publishing.
6. Ribeiro, J., Lima, R., Eckhardt, T., & Paiva, S. (2021). Robotic process automation and artificial intelligence in industry 4.0 – A literature review. *Procedia Computer Science, 181,* 51–58.
7. Syed, R., Suriadi, S., Adams, M., Bandara, W., Leemans, S. J., Ouyang, C., ... & Reijers, H. A. (2020). Robotic process automation: Contemporary themes and challenges. *Computers in Industry, 115,* 103162.
8. Van der Aalst, W. M., Bichler, M., & Heinzl, A. (2018). Robotic process automation. *Business & Information Systems Engineering, 60,* 269–272.
9. Ivančić, L., Suša Vugec, D., & Bosilj Vukšić, V. (2019). Robotic process automation: Systematic literature review. In *Business Process Management: Blockchain and Central and Eastern Europe Forum: BPM 2019 Blockchain and CEE Forum, Vienna, Austria, September 1–6, 2019, Proceedings 17* (pp. 280–295). Springer International Publishing.
10. Wajcman, J. (2017). Automation: Is it really different this time? *The British Journal of Sociology, 68*(1), 119–127.
11. Goldberg, K. (2011). What is automation? *IEEE Transactions on Automation Science and Engineering, 9*(1), 1–2.

12. Polo, M., Reales, P., Piattini, M., & Ebert, C. (2013). Test automation. *IEEE Software*, *30*(1), 84–89.
13. Fewster, M., & Graham, D. (1999). *Software Test Automation* (pp. 211–219). Reading: Addison-Wesley.
14. Thummalapenta, S., Sinha, S., Singhania, N., & Chandra, S. (2012, June). Automating test automation. In *2012 34th International Conference on Software Engineering (ICSE)* (pp. 881–891). IEEE.
15. Kasurinen, J., Taipale, O., & Smolander, K. (2010). Software test automation in practice: Empirical observations. *Advances in Software Engineering, 2010*.
16. Wiklund, K., Eldh, S., Sundmark, D., & Lundqvist, K. (2017). Impediments for software test automation: A systematic literature review. *Software Testing, Verification and Reliability*, *27*(8), e1639.

Chapter 3

Intelligent computing relating to cloud computing

Anjali Sharma, Nidhi Sindhwani,
Rohit Anand, and Rashmi Vashisth

INTRODUCTION

Organizations currently have new chances to make the most of the power of computerized reasoning and AI to work on the capacities of cloud-based applications and administrations on account of the combination of canny registering and distributed computing [1]. Man-made consciousness techniques, such as AI, normal language handling, and information examination, are utilized in wise processing to make PCs fit for undertakings that generally require human knowledge. The conveyance of processing assets by means of the web on a compensation for every utilization premise is known as distributed computing, conversely.

There are many benefits to incorporating savvy figuring techniques with distributed computing, including better asset usage, better versatility, and expanded security. AI calculations can be utilized to upgrade asset portion [2,3] and work on the proficiency of distributed computing frameworks, whereas regular language handling strategies can empower more normal and natural cooperations with cloud-based applications. Also, information examination can be utilized to separate experiences from enormous volumes of information put away in the cloud, empowering organizations to settle on additional educated choices.

In any case, the combination of clever figuring and distributed computing likewise presents critical difficulties, for example, the requirement for proficient asset on the board, information protection concerns, and the requirement for hearty safety efforts. The powerful idea of distributed computing, combined with the rising intricacy of canny processing procedures, requires new methodologies and answers to successfully influence the advantages of these innovations [4].

This chapter plans to give an exhaustive outline of distributed computing with regard to clever figuring. Particularly, this chapter will investigate the different methods and approaches utilized in shrewd figuring for distributed computing, break down the advantages and difficulties of coordinating savvy processing with distributed computing, and give bits of knowledge on the most proficient method to successfully use these advancements to

DOI: 10.1201/9781003479031-3

accomplish improved results. Moreover, this chapter will examine future headings for research around here, featuring the potential for savvy processing to change the scene of distributed computing.

Artificial intelligence

Man-made reasoning (simulated intelligence) [5,6] is a subfield of software engineering that spotlights on building clever machines that are fit for doing errands that normally require human knowledge, similar to discourse acknowledgment, language perception, navigation, and critical thinking. Computer-based intelligence intends to foster apparatuses that could not just do these assignments at any point but additionally help better at them after some time through learning and variation.

Man-made intelligence can be comprehensively grouped into two classes: tight or feeble artificial intelligence and general areas of strength for or. Limited or feeble artificial intelligence is intended to perform explicit errands, like facial acknowledgment or discourse interpretation, and isn't fit for general insight. Interestingly, general areas of strength are intended to be basically as wise as an individual and equipped for playing out any scholarly undertaking that a human would be able.

PC vision, mechanical technology, normal language handling, AI, and master frameworks are only a couple of the techniques and devices that make computer-based intelligence conceivable. Late improvements in computer-based intelligence have generally been driven by AI, which empowers machines to gain information and upgrade their exhibition over the long run. PC vision empowers machines to decipher visual information, whereas normal language handling empowers machines to understand and deal with human language.

Man-made intelligence has various applications in different fields, including medical services, money, transportation, and diversion. In medical services, for example, simulated intelligence can be utilized to dissect clinical pictures and anticipate sickness results, while in finance, man-made intelligence can be utilized for misrepresentation recognition and hazard the board [7]. Notwithstanding, simulated intelligence additionally presents huge difficulties and moral contemplations, for example, the potential for one-sided or prejudicial direction, the effect on work uprooting, and the potential for abuse in regions like independent weapons. In general, man-made intelligence addresses a quickly propelling field that holds huge potential for changing different enterprises and working on the personal satisfaction for individuals all over the planet [8–10].

Cloud computing

A technique for giving on-request admittance to shared processing assets over the web is called distributed computing. Processing power, stockpiling, and applications are a portion of these assets that can be provisioned and gotten to by clients on a compensation for each utilization premise. Organizations and people can get to registering assets utilizing the distributed computing model without buying, introducing, and keeping up with their own equipment or foundation [11].

Adaptability, versatility, and moderateness are properties of distributed computing. Clients can increase their registering assets or down depending on the situation, empowering them to deal with changes sought after for their applications or administrations. The utilization of virtualization, which empowers numerous virtual machines to run on a solitary actual machine, makes this versatility conceivable.

Public, private, and cross-breed mists are only a couple of instances of the different organization models used to convey distributed computing. Public mists are possessed and worked by outsider suppliers, for example, Amazon Web Administrations, Microsoft Sky blue, and Google Cloud Stage. Then again, confidential mists are restrictive to a solitary organization and are habitually arranged nearby or in a server farm [12]. Clients can profit from half-breed mists' mix of components from both public and confidential mists.

Distributed computing has various applications in different fields, including programming improvement, information examination, and man-made reasoning. Programming engineers can utilize distributed computing to fabricate and convey applications all the more rapidly and effectively, while information investigators can use distributed computing to store and deal with huge volumes of information. With cloud suppliers presently offering computer-based intelligence stages and administrations that let engineers make and use artificial intelligence models, simulated intelligence specifically has taken huge steps in the time of distributed computing [13].

Major Components of Intelligent Cloud are shown in Figure 3.1.

Effective computing

The objective of the interdisciplinary field of study known as full of feeling registering is to make computational frameworks that can perceive, understand, and respond to human feelings. Full of feeling figuring expects to foster clever frameworks that collaborate with individuals in a more natural [14] way by distinguishing and responding to profound prompts in verbal correspondence.

Emotional processing draws on a scope of fields, including brain research, neuroscience, software engineering, and computerized reasoning. Its point is to foster frameworks that can figure out human feelings, such as euphoria, outrage, dread, and misery, and answer suitably, whether it is to offer help, solace, or consolation.

Various advancements, for example, AI, PC vision, discourse acknowledgment, and regular language handling, empower full of feeling processing.

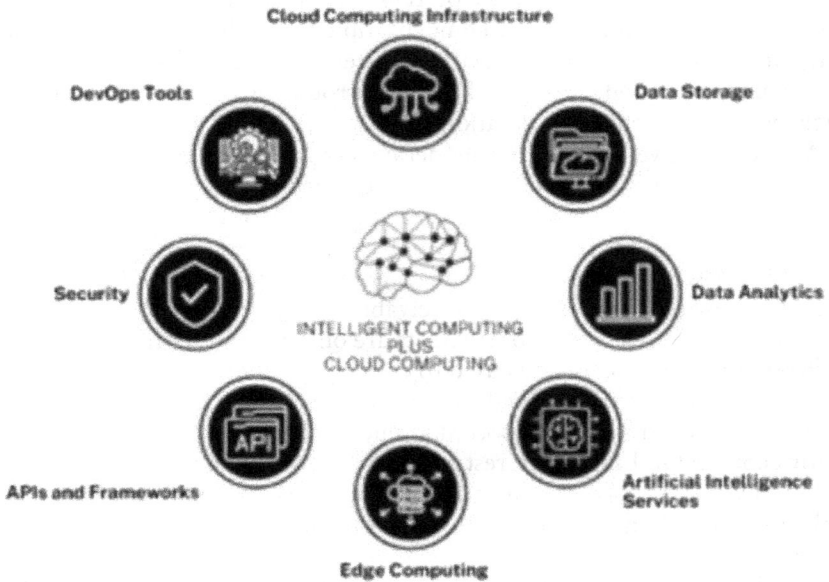

Figure 3.1 Major components of Intelligent Cloud.

With the guidance of these innovations, PCs are currently ready to examine different information sources, including physiological signs, looks, discourse examples, and discourse designs, to find out the close-to-home condition of the client [15].

Emotional registering has various applications in different fields, such as medical services, schooling, amusement, and promoting. In medical care, full of feeling registering can be utilized to analyze and treat emotional wellness issues, while in training, it tends to be utilized to give customized opportunities for growth. Full of feeling processing can be utilized in media outlets to deliver more vivid and sincerely convincing encounters, and it can likewise be utilized in showcasing to perceive and better address the issues and inclinations of purchasers.

In any case, emotional processing likewise raises huge moral worries, like security, predisposition, and the potential for abuse. For example, the utilization of full of feeling registering in observation and policing huge worries about protection and common freedoms. In general, it addresses an astonishing and quickly propelling field that holds incredible potential for further developing human-PC collaboration and improving comprehension we might interpret human feelings [16].

LITERATURE REVIEW

The two prospering innovations of distributed computing and wise figuring are hugely affecting numerous areas of current registering. With an end goal to enhance the proficiency and execution of cloud-based administrations, there has been a flooding energy in joining keen processing approaches with distributed computing as of late.

This writing survey will break down a portion of the key examinations that dig into the relationship between distributed computing and smart processing.

"Intelligent Computing in Cloud Computing" by V. Vijayarani notwithstanding P. Deivendran: To update the well-being and viability of distributed computing, this paper examines the fuse of astute processing methods like brain organizations, hereditary calculations, and fluffy rationale [17].

"An Intelligent Computing Model for Cloud Computing Resource Allocation" by Q. Zhang and Y. Chen: This paper recommends a hereditary calculation-based canny figuring model for designating distributed computing assets. The plan is designed to expand the distribution of materials across different cloud-based programs in accordance with their particular necessities and highlights [18].

"Cloud Computing with Intelligent Computing Techniques for Big Data Analytics" by S. M. Sarwar and A. Ullah: This article analyses the use of state-of-the-art figuring methods, for example, AI and computer-based intelligence, in cloud-based enormous information examination. The creators propose a design to address the difficulties of large information investigation by consolidating distributed computing and high-level figuring [19].

"Intelligent Cloud Computing: A Review" by S. S. Bedi and A. Sharma: This piece gives a thorough audit of the ongoing scene of distributed computing and knowledge. It covers different kinds of man-made brainpower, AI, and profound discoveries that are utilized in cloud innovation [20].

"Intelligent Edge Computing for Cloud-Based IoT Applications" by A. H. Alissa and M. Z. Al-Fedaghi: This study evaluates the connection between distributed computing and shrewd edge processing for IoT applications. The creators' clever edge processing model, which carries out AI and profound learning strategies, is utilized to improve the presentation and energy productivity of IoT applications running in the cloud [21].

Subsequently, mixing progressed registering strategies into distributed computing can essentially support its highlights, well-being, and efficiency. The points shrouded in this study can be utilized as an establishment for additional exploration around here.

METHODOLOGY

The expression "astute registering" suggests the use of computational strategies, for instance, AI, normal language handling, PC vision, and mechanical technology, to copy or update human discernment. On the other hand, distributed computing alludes to the contribution of processing resources and administrations over the web [22].

The following advances are a piece of the methodology for intelligent computing with regard to cloud computing:

1. **Problem Identification:** The initial step in intelligent computing is to precisely recognize the issue or chance that computational strategies can be utilized to tackle.
2. **Data Collection and Preparation:** After recognizing an issue, the subsequent stage is to find and set up the data that will be utilized in the investigation. This incorporates recognizing pertinent data sources, gathering and purifying the data, and putting it in a configuration that will be helpful for examination.
3. **Cloud Infrastructure Setup:** This includes choosing the perfect cloud computing service, arranging the essential facilities and materials, and ensuring the structure is adjustable and can manage fluctuating scales of activity.
4. **Model Development:** The subsequent stage is to form the computational model that will be applied to the problem. This includes opting for the optimal computer vision, natural language processing, or machine learning technique, preparing the model utilizing the pre-made data, and examining the model's competency.
5. **Model Deployment:** The model must be uploaded to the web-based platform after it has been constructed and educated. This requires dispatching the model, settling on an acceptable release plan, and guaranteeing the model is adaptable and fit for dealing with various traffic volumes. Prompt
6. **Real-Time Data Processing:** After the model is utilized, it tends to be utilized to process data quickly. This includes handling data in real time and giving investigations, conjectures, or proposals in view of the model's decisions.
7. **Model Monitoring:** After the model has been utilized, it should be checked to ensure it is working as expected. This involves monitoring the model's performance, recognizing any issues or errors, and instituting any necessary modifications to the model or cloud setting.
8. **Model Optimization:** This step is pivotal to ensuring that the model is delivering accurate outcomes. Examining the model's performance metrics, discovering prospective regions for growth, and effecting the vital changes to the model or the data flow are all components of this procedure.

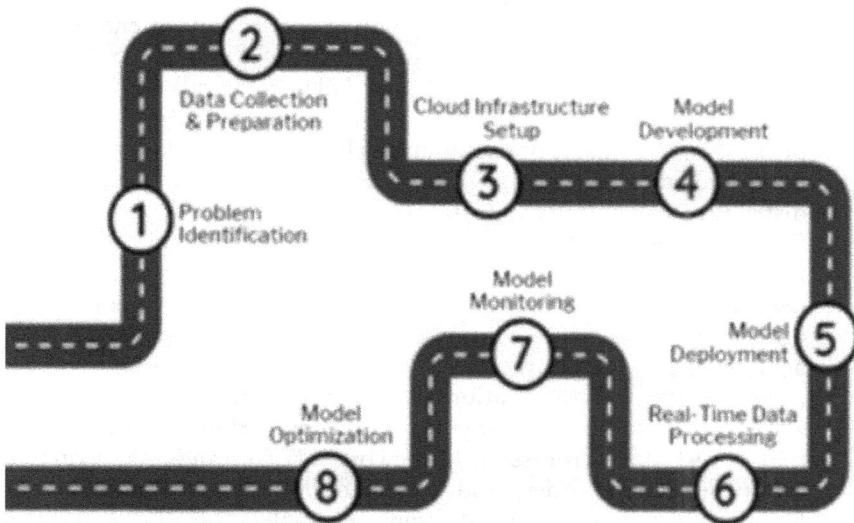

Figure 3.2 Integration of intelligent computing and cloud computing.

Applying this approach, companies can create and carry out smart computing solutions that can make a tangible difference to their bottom line by taking advantage of cloud computing resources. It gives companies access to the computing power and storage space necessary to process huge amounts of information, construct complex computational models, and present timely insights and advice.

The integration of intelligent Computing and Cloud Computing is shown in Figure 3.2.

Techniques

Intelligent computing techniques are used extensively in cloud computing applications to process and analyze large volumes of data. Some of the most common techniques used in intelligent computing relating to cloud computing include:

1. **Machine Learning:** AI is a part of man-made consciousness that enables PCs to gain from information and pursue forecasts or choices without being unequivocally modified. To process and dissect huge datasets and simplify them to recognize examples and experiences, AI calculations are generally utilized in distributed computing applications [23].
2. **Deep Learning:** Profound learning is a part of AI that cycles and examines information utilizing brain organizations. For assignments

like picture acknowledgment, discourse acknowledgment, and normal language handling, distributed computing applications oftentimes utilize profound learning calculations.

3. **Natural Language Handling**: Regular language handling (NLP) is a wise figuring strategy that empowers machines to fathom and decipher human language. For assignments like feeling examination, chatbots, and menial helpers, NLP calculations are broadly utilized in distributed computing applications.

4. **Computer Vision**: Machine investigation and understanding of visual information spread the word about conceivable by clever processing methods such as PC vision. Applications for distributed computing often use PC vision calculations to perform undertakings like item acknowledgment, picture order, and facial acknowledgment [24–26].

5. **Artificial Brain Organizations**: Roused by the construction and activity of natural brain organizations, counterfeit brain organizations (ANN) are processing frameworks. For errands like example acknowledgment, order, and expectation, ANN calculations are broadly utilized in distributed computing applications [27].

Approaches

There are several approaches used in intelligent computing relating to cloud computing, which are described below:

1. **Centralized Methodology**: In this methodology, every one of the information is gathered and handled in a focal area inside the cloud. This approach is in many cases utilized in applications where there is a lot of information to be handled, and the handling should be done rapidly. The incorporated methodology can be more viable than a decentralized procedure since it requires less information to be sent over the organization [28].

2. **Decentralized Methodology**: In this methodology, the information is gathered and handled in various areas inside the cloud. This approach is much of the time utilized in applications where the information is conveyed across numerous areas, or where the handling should be done locally to diminish dormancy. The decentralized methodology can be more issue-lenient than a concentrated methodology since it decreases the gamble of a weak link.

3. **Hybrid Methodology**: In this methodology, a mix of unified and decentralized handling is utilized. This approach is much of the time utilized in applications where some handling should be done rapidly, while other handling should be possible all the more leisurely. The half and half methodology can be more adaptable than either the concentrated or decentralized approach since it can adjust to changing handling necessities.

4. **Edge Registering Approach:** This strategy places information handling nearer to the information source, at the organization's edge. This technique is much of the time utilized in programs where low idleness is fundamental, such as modern mechanization or independent vehicles. Since less information should be communicated over the organization while utilizing the edge figuring approach, it very well might be more successful than brought together or decentralized approaches [29].

5. **Federated Learning Approach:** With this technique, different organization hubs or gadgets cooperate to prepare an AI model without moving crude information between them. This technique is as often as possible applied in fields like medical care and money where information security is fundamental.

By and large, the methodology utilized in wise registering connecting with distributed computing relies upon the particular prerequisites of the application, such as how much information, handling pace, and information security necessities.

Latest advances

There have been several recent advances in intelligent computing related to cloud computing, including:

1. **Machine Learning-as-a-Service (MLaaS):** This alludes to the cloud-based conveyance of AI administrations, permitting clients to handily get to and use AI calculations without the requirement for specific equipment or programming. AI models can be prepared, conveyed, and overseen in the cloud with the assistance of suppliers like Amazon Web Administrations, Microsoft Purplish blue, and Google Cloud.

2. **Serverless Figuring:** This is another sort of distributed computing where the specialist organization controls the foundation and naturally disseminates assets to run and scale applications. Because of its ability to oversee complex errands, improve asset usage, and lower costs, serverless processing has become increasingly famous.

3. Automating redundant and exhausting errands utilizing man-made brainpower (simulated intelligence) and AI (ML) advances is known as insightful computerization [30]. Intelligent automation solutions are being offered by cloud providers more frequently now, enabling companies to automate operations like supply chain management, customer service, and human resources.

4. Edge computing is the process of processing and analyzing data locally, as opposed to sending it to a centralized cloud data center, usually on the source or endpoint device. This approach is gaining popularity due to the increasing need for real-time data processing and low latency in applications such as IoT, autonomous vehicles, and AR/VR [31].

Overall, these advancements are helping to drive the adoption of intelligent computing in cloud computing, enabling organizations to derive greater value from their data and improve business outcomes.

Ethics

Smart figuring, which utilizations AI (ML) and man-made reasoning (computer-based intelligence) calculations, has huge moral repercussions for distributed computing. These are a couple of the main moral inquiries encompassing distributed computing's clever figuring [32]:

1. **Data Privacy**: To build models for intelligent computing, large amounts of data are frequently processed and stored in cloud computing environments. The collection and processing of data for intelligent computing must be done with the explicit consent of the users, and the data must be safeguarded against improper access, use, or disclosure [33].
2. **Transparency**: Machine learning algorithms used in intelligent computing may be cryptic and challenging to comprehend, making it challenging to detect and address bias or other ethical concerns. It is crucial to make machine learning algorithms transparent and understandable so that users can comprehend how they operate and how they arrive at decisions.
3. **Accountability**: Because the decisions made by intelligent computing systems can have a big impact on people's lives and society as a whole, it's crucial to make sure that these systems are held accountable for their actions.
4. **Governance and Regulation**: The utilization of keen registering in distributed computing raises huge issues with respect to government and regulation [34]. To guarantee that canny figuring frameworks are created, sent, and utilized in a moral and dependable way, ensuring that the essential regulations and norms are in place is urgent.

In general, there are many different ethical issues that arise with intelligent computing and cloud computing. To ensure that their use of intelligent computing in cloud computing is responsible and ethical, it is crucial for organizations to be aware of these factors and to take proactive measures.

BENEFITS, CHALLENGES AND APPLICATIONS

The term "intelligent computing" refers to a computer system's capacity for learning, reasoning, and decision-making based on data inputs. Intelligent computing can have many uses and advantages when combined with cloud computing, but it also has some drawbacks.

Benefits

Intelligent computing and cloud computing have several benefits when used together. The following are some of the benefits of intelligent computing relating to cloud computing:

1. **Scalability:** Intelligent computing applications often require significant computational resources to process large datasets. Cloud computing provides an elastic and scalable infrastructure that can allocate resources dynamically to meet the changing needs of these applications.
2. **Reduced Infrastructure Costs:** By utilizing distributed computing, associations can keep away from the significant expenses of building and keeping up with on-premise registering foundation. Distributed computing suppliers can offer admittance to strong equipment and programming devices at a lower cost than associations can accomplish all alone [35].
3. **Improved Data Processing Capabilities:** Intelligent computing algorithms, such as machine learning and artificial intelligence, require significant amounts of data to learn and make decisions. By storing data in the cloud, organizations can take advantage of cloud computing's advanced data processing capabilities, such as distributed computing and parallel processing.
4. **Increased Collaboration:** Intelligent computing applications in the cloud can enable teams to work collaboratively on projects, regardless of their physical location. Cloud computing also makes it easier to share data, code, and models across teams, accelerating the pace of innovation.
5. **Enhanced Security:** High-level safety efforts are given by distributed computing suppliers to shield information and applications, including encryption and access controls. This can assist associations with meeting consistence prerequisites and moderate the dangers of digital assaults [36].

In summary, intelligent computing and cloud computing together provide organizations with a scalable, cost-effective, and secure infrastructure that can significantly enhance their data processing capabilities, leading to better decision-making and increased innovation.

Challenges

To fully realize its potential, intelligent computing in relation to cloud computing must overcome a number of obstacles. Some of the main difficulties are listed below [37]:

1. **Secure Data:** These are issues that are turning out to be increasingly squeezing as distributed computing turns out to be more common. Large data sets are necessary for intelligent computing systems, and

processing and storing these data in the cloud can be extremely risky in terms of security. It is crucial to ensure that data is protected at all stages, from storage to processing and transmission.

2. **Interoperability and Integration:** Intelligent computing systems often rely on multiple technologies and platforms, making it challenging to ensure interoperability and integration. Different platforms and technologies may use different data formats, communication protocols, and APIs, making it difficult to exchange data and integrate different systems.

3. **Resource Allocation and Management:** Intelligent computing systems require significant computing resources to function, and it is crucial to ensure that these resources are allocated and managed effectively [38]. Cloud computing providers must ensure that their systems can handle the demand for computing resources, while also ensuring that these resources are used efficiently.

4. **Ethical Concerns:** Intelligent computing systems can raise ethical concerns, particularly in areas such as healthcare, finance, and public policy. For example, there may be concerns about bias, transparency, and accountability in decision-making processes. It is crucial to make sure that these systems are created and used ethically, with the proper oversight and regulation.

5. **Cost and Scalability:** Building and deploying intelligent computing systems can be costly, particularly for smaller organizations. It is essential to ensure that these systems are scalable and affordable, enabling organizations of all sizes to benefit from intelligent computing.

6. **Performance and Latency:** As intelligent computing systems become more sophisticated, they require increasingly powerful computing resources. Ensuring that these systems can perform at scale, while also maintaining low latency and high performance, is a significant challenge [39].

Addressing these challenges requires a coordinated effort from researchers, cloud providers, policymakers, and other stakeholders. By working together to overcome these challenges, we can ensure that intelligent computing in relation to cloud computing continues to evolve and improve, enabling organizations to achieve new levels of efficiency, productivity, and innovation.

Applications

Cloud computing can benefit from a variety of applications for intelligent computing [40] to increase its functionality and effectiveness. Here are a few illustrations:

1. **Predictive Analytics:** Predictive models that can be used to examine a lot of data in the cloud can be created using intelligent computing. This can help in spotting potential problems before they happen and in spotting new business opportunities.

2. **Natural Language Processing (NLP):** NLP can be used in the cloud to create smart chatbots that can assist businesses in interacting with their customers. These chatbots can be programmed to handle transactions, respond to customer questions, and provide assistance.
3. **Machine Learning:** Cloud-based information can be utilized to prepare AI calculations to track down examples and patterns in information. As a result, businesses may be better able to make decisions based on data insights.
4. **Data Mining:** To uncover buried patterns and trends in sizable datasets stored in the cloud, intelligent computing can be used. Making data-driven decisions based on this information can help businesses better understand their customers.
5. **Cloud Security:** Security models that can be used to safeguard data in the cloud can be developed using intelligent computing. Data leaks, hacking, and other cyberattacks might be avoided as a result of this.
6. **Resource Management:** Intelligent computing can be applied to enhance cloud resource management. This could aid companies in more cost-effectively managing their cloud resources [41,42]

In conclusion, a variety of applications for intelligent computing can be used to enhance the functionality and effectiveness of cloud computing.

CONCLUSION

Taking everything into account, smart registering and distributed computing are two quickly developing fields that can possibly alter the manner in which we approach information handling and stockpiling. The term keen processing depicts the utilization of AI (ML) and man-made consciousness (man-made intelligence) calculations to empower PCs to perform choices and undertakings that customarily require human knowledge [43]. Conversely, distributed computing depicts the act of putting away and getting information and programming over the web rather than a nearby PC.

Keen processing and distributed computing are firmly connected, with numerous clever registering applications depending on the distributed computing framework to work. One huge benefit of utilizing distributed computing for insightful registering is that it gives a versatile and financially savvy method for handling a lot of information, which is fundamental for some computer-based intelligence and ML applications.

In the context of cloud computing, the review paper examined various applications of intelligent computing, such as NLP, computer vision, and predictive analytics. The paper also examined the challenges associated with using intelligent computing in the cloud, such as privacy and security concerns, and the need for specialized hardware to handle the computationally intensive tasks involved in AI and ML [44].

In spite of these difficulties, the paper reasoned that the combination of clever registering with distributed computing can possibly change numerous enterprises, including medical care, money, and transportation. With the proceeded with the advancement of computer-based intelligence and ML calculations, as well as upgrades in distributed computing framework, almost certainly, we will see considerably more creative and significant applications later on. Notwithstanding, it is likewise crucial to address the moral and social ramifications of these advances to guarantee that they are utilized in a capable and valuable manner for society in general.

FUTURE SCOPE AND RECOMMENDATIONS

The future scope of intelligent computing in relation to cloud computing is vast, with many exciting opportunities for innovation and growth. Here are some potential areas of focus for the future:

1. **Edge Computing:** As more intelligent applications require real-time processing and low latency; edge computing is becoming an essential component of intelligent computing in the cloud. Edge computing enables data to be processed nearer to the point of origin, decreasing latency and enhancing performance. Future developments in edge computing are likely to focus on integrating it more closely with cloud computing, enabling more seamless integration and greater scalability [45].
2. **Hybrid Cloud Environments:** Mixture cloud conditions are making progress as organizations hope to find some kind of harmony between the benefits of public distributed computing and the security and control of private distributed computing. Shrewd processing will assume a basic part in empowering consistent reconciliation among public and confidential mists, permitting associations to use the best-case scenario.
3. **More Sophisticated Intelligent Services:** Cloud providers are continually improving their intelligent services, offering more sophisticated and powerful capabilities such as NLP, image recognition, and predictive analytics. As the services become more accessible and affordable, we can expect to see a proliferation of intelligent applications across a wide range of industries.
4. **Continued Growth in AI and ML:** As AI and ML continue to evolve and improve, they will play an increasingly important role in intelligent computing. Future developments in these fields are likely to focus on improving the accuracy and reliability of intelligent systems, enabling more sophisticated applications and better outcomes [46].

Overall, the future of intelligent computing in relation to cloud computing is bright, with many exciting opportunities for innovation and growth. As these technologies continue to evolve and improve, we can expect to see intelligent applications become more widespread and more sophisticated, enabling organizations to achieve new levels of efficiency, and productivity.

Recommendations

1. **Invest in AI and Machine Learning:** Firms should allocate funds to these advancements to amplify the automation and acumen of their cloud services. This will equip them to give their customers more useful and personalized services [47].
2. **Focus on Security and Privacy:** When transitioning to the cloud, organizations should give utmost attention to the security and confidentiality of their data and applications. This can be achieved by employing sophisticated security systems that are able to detect and prevent digital dangers.
3. **Explore Edge Computing:** Businesses should explore the utilization of edge computing for rapid data processing and assessment, particularly for applications like the Internet of Things (IoT) gadgets and self-driving cars.
4. **Collaborate with Cloud Providers:** To gain access to the most recent services and technologies for intelligent computing, companies should work together with cloud service providers. This will permit them to sustain their edge and afford their customers superior worth [48].

REFERENCES

1. Guo, F., Du, Q., Yang, J., "Design and implementation of an intelligent cross platform IaaS", *Advanced Materials Research*, 2014.
2. Hosseini, P., Taheri, S., Akhavan, J., Razban, A., "Privacy-preserving federated learning: Application to behind-the-meter solar photovoltaic generation forecasting", *Energy Conversion and Management*, 2023
3. Buyya, R., *Cloud Computing Principles and Paradigms*, John Wiley & Sons Ltd, 2014.
4. Erl, T., Puttini, R., Mahmood, Z., *Cloud Computing: Concepts, Technology & Architecture*, Prentice Hall, May 2, 2013
5. Lunger, G.F., Stubblefield, W.A., *Artificial Intelligence – Structures and Strategies for Complex Problem Solving*, Benjamin-Cummings, Albuquerqe, ISBN 0-8053-4780-1, 1993.
6. Russell, S.J., Norvig, P., *Artificial Intelligence: A Modern Approach*, Prentice Hall, 2003.
7. Rich, E., Knight, K., *Artificial Intelligence*, McGraw Hills Inc., 2006.

8. Sindhwani, N., Anand, R., Vashisth, R., Chauhan, S., Talukdar, V., Dhabliya, D., Thingspeak-based environmental monitoring system using IoT. In *2022 Seventh International Conference on Parallel, Distributed and Grid Computing (PDGC)* (pp. 675–680), IEEE, November 2022.

9. Singh, P., Kaiwartya, O., Sindhwani, N., Jain, V., Anand, R. (Eds.), *Networking Technologies in Smart Healthcare: Innovations and Analytical Approaches*, CRC Press, 2022.

10. Meivel, S., Sindhwani, N., Anand, R., Pandey, D., Alnuaim, A.A., Altheneyan, A.S., … Lelisho, M.E., Mask detection and social distance identification using internet of things and faster R-CNN algorithm, *Computational Intelligence and Neuroscience*, 2022.

11. *Innovations and Advances in Computing, Informatics, Systems Sciences, Networking and Engineering*, Springer Science and Business Media LLC, 2015

12. Goscinski, A., *Cloud Computing: Concepts, Technologies and Architectures*, IGI Global, 2012.

13. Babura an, R., The rising cloud storage market opportunity strengthens vendors, "InfoTech", It.tmcnet.com, August 24, 2011. Retrieved 2011-12-02.

14. Encrypted storage and key management for the cloud.

15. Bechtolsheim, A. Chairman & Co-founder, Arista networks, November 12, 2008.

16. Picard, R.W., *Affective Computing*, MIT Press, 2000.

17. Vijayarani, V., Deivendran, P., *Intelligent Computing in Cloud Computing*.

18. Zhang, Q., Chen, Y., *An Intelligent Computing Model for Cloud Computing Resource Allocation*.

19. Sarwar, S.M., Ullah, A., *Cloud Computing with Intelligent Computing Techniques for Big Data Analytics*.

20. Bedi, S.S., Sharma, A., *Intelligent Cloud Computing: A Review*.

21. Alissa, A.H., Al-Fedaghi, M.Z., *Intelligent Edge Computing for Cloud-Based IoT Applications*.

22. Hyde, A. Dean, *The Future of Artificial Intelligence*, September 28, 2010.

23. Goodfellow, I., Bengio, Y., Courville, A., *Deep Learning*, MIT Press, 2016

24. Sindhwani, N., Anand, R., Meivel, S., Shukla, R., Yadav, M. P., Yadav, V., Performance analysis of deep neural networks using computer vision, *EAI Endorsed Transactions on Industrial Networks and Intelligent Systems*, 8(29), e3–e3, 2021.

25. Jain, S., Sindhwani, N., Anand, R., Kannan, R., COVID detection using chest X-ray and transfer learning. In *Intelligent Systems Design and Applications: 21st International Conference on Intelligent Systems Design and Applications (ISDA 2021)* held during December 13–15, 2021 (pp. 933–943), Cham: Springer International Publishing, March 2022.

26. Lalitha Kumari, P., Das, S., Kannadasan, B., Sampathila, N., Saravanakumar, C., Anand, R., Gupta, A. Methodology for classifying objects in high-resolution optical images using deep learning techniques. In *Advances in Signal Processing, Embedded Systems and IoT: Proceedings of Seventh ICMEET-2022* (pp. 619–629), Singapore: Springer Nature Singapore.

27. Deng, L., Hinton, G., Kingsbury, B., New types of deep neural network learning for speech recognition and related applications: An overview. In *Acoustics, Speech and Signal Processing (ICASSP), 2013 IEEE International Conference*, IEEE, 2013.
28. *Principles of Internet of Things (IoT) Ecosystem: Insight Paradigm*, Springer Science and Business Media LLC, 2020
29. Kaur, J., Sindhwani, N., Anand, R., Pandey, D., Implementation of IoT in various domains. In *IoT Based Smart Applications* (pp. 165–178), Cham: Springer International Publishing, 2022.
30. Tadapaneni, N.R., Cloud computing: Opportunities and challenges, *International Journal of Technical Research and Applications*. SSRN Electronic Journal.10.2139/ssrn.3563342, 2018
31. Sindhwani, N., Maurya, V.P., Patel, A., Yadav, R.K., Krishna, S., Anand, R., Implementation of intelligent plantation system using virtual IoT. *Internet of Things and Its Applications*, 305–322.
32. Ryan, P.S., Merchant, R., Falvey, S., Regulation of the cloud in India (July 30, 2011), *Journal of Internet Law*, 15(4), 7, October 2011, Available at SSRN.
33. Anand, R., Shrivastava, G., Gupta, S., Peng, S.L., Sindhwani, N., Audio watermarking with reduced number of random samples. In *Handbook of Research on Network Forensics and Analysis Techniques* (pp. 372–394), IGI Global, 2018.
34. Ostrich, K., Converged infrastructure, CTO Forum, Thectoforum.com, November 15, 2010. Retrieved 2011-12-02.
35. Jin, H., et al., Tools and technologies for building clouds. In *Cloud Computing* (pp. 3–20), London: Springer, 2010.
36. Schauhan, Cs., A shift from cloud computing model to fog computing, *Journal of Applied Computing*, 2016.
37. Service-oriented computing and cloud computing: Challenges and opportunities, *IEEE Internet Computing*. Retrieved 2010-12-04.
38. Arora, I., Gupta, A., Cloud databases – A paradigm shift in databases, *IJCSI International Journal of Computer Science Issues*, 9(4, No 3), July 2, 2012.
39. Renta, T., The role of IoT and cloud computing in health monitoring systems. In *IEEE 19th International Conference on Bioinformatics and Bioengineering*, 2018.
40. Daniel M., *Databases in the Cloud*, Rapperswil, 2010.
41. Taigman, Y., et al., Deep face: Closing the gap to human-level performance in face verification. In *Proceedings of the IEEE Conference on Computer Vision and Pattern Recognition*, 2014.
42. Poole, D., Mackworth, A., Goebel, R., *Computational Intelligence*, Oxford: Oxford University Press, 1998.
43. Bengio, Y., Learning deep architectures for AI, *Machine Learning*, 2(1), 1–127, 2009.
44. Mills, E., Cloud computing security forecast: Clear skies, *CNET*, January 27, 2009.
45. *The NIST Definition of Cloud Computing*, National Institute of Science and Technology, July 24, 2011.

46. Anand, R., Sindhwani, N., Juneja, S., Cognitive Internet of Things, its applications, and its challenges: A survey. In *Harnessing the Internet of Things (IoT) for a Hyper-Connected Smart World* (pp. 91–113), Apple Academic Press, 2022.
47. Sindhwani, N., Rana, A., Chaudhary, A., Breast cancer detection using machine learning algorithms. In *2021 9th International Conference on Reliability, Infocom Technologies and Optimization (Trends and Future Directions) (ICRITO)* (pp. 1–5), IEEE, September, 2021.
48. Encrypted storage and key management for the cloud. Cryptoclarity.com, July 30, 2009.

Chapter 4

Communication protocols used for industrial automation

Richa Bansal and Anil Kumar Dubey

Various communication protocols are necessary for industrial automation in order for devices and systems to exchange data and control signals [20, 21, 24]. These procedures are essential for maintaining reliable and effective operation in industrial settings. Several popular industrial automation communication [1, 3, 10, 15–17, 22] protocols are listed below:

1. **Modbus:** A popular serial communication protocol is called Modbus. It is a common choice for attaching field devices like sensors and actuators to programmable logic controllers (PLCs) and other industrial controls since it is straightforward and open [4,5, 6, 19, 22].
2. **Profibus:** Field devices and control systems can communicate using the fieldbus protocol [9] Profibus. It is available in a variety of forms, such as Profibus DP for high-speed communication and Profibus PA for process automation [7, 8].
3. **DeviceNet:** For connecting and managing industrial devices, DeviceNet is a network protocol used in industrial automation. It frequently appears in I/O and motor control applications.
4. **Ethernet/IP:** An industrial Ethernet standard called Ethernet/IP enables easy communication between gadgets made by various vendors [11, 12, 13, 23]. It is based on Ethernet protocol standard and employs Common Industrial Protocol (CIP).
5. **Profinet:** Another industrial Ethernet protocol, called Profinet, is mostly utilized in Europe. It provides fast connection and works with several Ethernet standards.
6. **CAN (Controller Area Network):** Automotive and industrial applications employ the sturdy and dependable serial communication standard known as CAN. It is popular in distributed control systems because of its fault tolerance.
7. **Foundation Fieldbus:** A widely used protocol for process control applications is called Fieldbus. It is intended for use in process industries as a real-time control and monitoring system for field devices.

DOI: 10.1201/9781003479031-4

8. **HART (Highway Addressable Remote Transducer)**: Process industries frequently use HART, a hybrid analog/digital communication protocol, with smart field equipment. Along with the analog stream, more digital data may be transmitted [18].

9. **Modbus TCP/IP**: A modification to the Modbus protocol called Modbus TCP/IP enables Modbus communication over Ethernet networks. It is frequently utilized to link PLCs and other hardware to SCADA systems.

10. **OPC (OLE for Process Control)**: OPC is a set of standards and specifications that enable interoperability between various industrial automation devices and software applications but is not a communication protocol in and of itself.

11. **EtherCAT**: An Ethernet-based fieldbus system called EtherCAT is renowned for its real-time and high-speed communication capabilities. It is frequently employed in applications that call for rapid data interchange and exact synchronization.

12. **CANopen**: A higher-layer protocol built on the CAN physical layer is known as CANopen. It is utilized for device-to-device communication in a variety of applications, such as automation and motion control.

13. **Modbus RTU**: Binary encoding and serial (RS-232 or RS-485) transmission are used in the Modbus RTU protocol, a variation of the Modbus standard. It is frequently utilized in applications that call for lengthy cable runs.

14. **CC-Link**: A series of network protocols known as CC-Link is mostly utilized in Asia for industrial automation. To accommodate various applications, it provides a range of speed and topology options.

15. **BACnet**: BACnet is sometimes utilized in industrial settings for HVAC management and monitoring, despite being more frequently connected with building automation systems (BAS).

MODBUS

A PLC, other industrial equipment, and other devices in a supervisory control and data acquisition (SCADA) system frequently communicate with one another via the Modbus communication protocol, which is widely used in industrial automation. Here are a few Modbus essentials:

History: Modbus was created in 1979 as a serial communication protocol by Modicon, which is now a part of Schneider Electric. Since then, it has changed and grown to accommodate a variety of communication channels, such as Modbus RTU (serial), Modbus ASCII (serial), and Modbus TCP/IP (Ethernet-based).

Serial versions

Modbus RTU: A binary protocol called Modbus RTU (Remote Terminal Unit) makes advantage of serial communication (usually RS-232 or RS-485). It is renowned for its effectiveness and is frequently used for long-distance communication in industrial settings.

Modbus ASCII: Another serial version of Modbus is Modbus ASCII. Although it is human understandable and uses ASCII letters to represent data, it is less effective than Modbus RTU.

Ethernet-based version

Modbus TCP/IP: The Modbus protocol has been extended to run over Ethernet networks with Modbus TCP/IP. Modern industrial automation systems frequently employ it to speed up data transfer.

Master-slave architecture

Typically, master-slave architecture is used with Modbus. One or more slave devices (such as sensors, actuators, or other control devices) receive data requests or commands from the master device (such as a PLC or SCADA system), which starts communication. The slaves comply with the master's instructions.

Data Types: Input registers (16-bit integer data), holding registers (16-bit integer data), discrete inputs (binary), coils (binary outputs), and other data formats are all supported by Modbus. Due to its adaptability, it can manage a variety of industrial data.

Addressing: To help network administrators identify Modbus devices, they are often given distinctive addresses. In order for the master to know which slave device to connect with, addressing is necessary.

Function Codes: The sort of operation that will be carried out is specified using function codes in Modbus. In contrast, function code 5 is used to write a single coil (output) and function code 3 is utilized to read analog data from a holding register.

Error Handling: Error-checking techniques are built into Modbus to guarantee data integrity. Cyclic Redundancy Check (CRC) is used for error detection and correction.

Open Standard: Being an open protocol, Modbus is independent of any one maker or supplier. Its extensive adoption and compatibility with products from many manufacturers are due to its openness.

Applications: Numerous industrial applications, such as manufacturing, energy management, building automation, and others, use Modbus. Applications where dependable device-to-device communication is required are ideally suited for it.

Interoperability: It is comparatively simple to connect devices from several manufacturers into a single industrial control system thanks to Modbus' open design and widespread support.

Limitations: Although Modbus is a strong and dependable protocol, it might not be the best option for applications requiring exceptionally high speeds or a short response time. Other industrial Ethernet protocols like EtherCAT or Profinet may be preferred in certain situations.

PROFIBUS

In the manufacturing and process control industries, automation and field equipment are typically connected to and controlled via PROFIBUS (Process Field Bus), a popular industrial communication standard. It was created in the late 1980s and has since undergone various iterations, the most popular of which are PROFIBUS DP (Decentralized Peripherals) and PROFIBUS PA (Process Automation). Here are some salient characteristics and specifics of the PROFIBUS protocol:

Communication types

- **PROFIBUS DP:** It is used to connect sensors, actuators, and other field devices to PLCs and other control systems. This is the most popular form. Fast data transfer is supported for real-time control applications.
- **PROFIBUS PA:** This type was created especially for applications involving process automation, like those found in the chemical and petrochemical industries. Although it functions more slowly than PROFIBUS DP, it is designed for dangerous environments.

Physical Layers: Electrical signaling in PROFIBUS is accomplished using Manchester encoding and RS-485 (for PROFIBUS DP). It uses a two-wire, current-loop method of communication with field devices for PROFIBUS PA.

Topology: Both bus and star topologies are supported by PROFIBUS DP. Devices are daisy-chained on a single communication line in a bus architecture. Each device in a star topology is directly connected to the main hub.

Typically, PROFIBUS PA uses a two-wire, bus-like topology.

Data Transmission: A master device (such as PLC) controls and communicates with numerous slave devices (field devices) using the master-slave communication mechanism that PROFIBUS employs. Both cyclic and acyclic communication are supported. Process data, such as sensor readings, and parameterization data, such as device setup, are exchanged.

Speed: Depending on the cable length and baud rate, PROFIBUS DP can handle data transfer speeds between 9.6 Kbps and 12 Mbps. Typically, PROFIBUS PA works at slower speeds of approximately 31.25 Kbps.

Diagnosis and Maintenance: To help identify flaws and problems in the communication network, PROFIBUS offers diagnostic functions. It facilitates device parameterization and online configuration, making it simpler to replace or add additional field devices.

International Standards: As a result of being standardized under IEC 61158 and IEC 61784, PROFIBUS is a communication protocol that is known throughout the world.

Compatibility: Numerous manufacturers of automation equipment support PROFIBUS and it has gained widespread use across numerous industries.

DEVICENET

DeviceNet is a network technology and industrial communication protocol that is frequently employed in manufacturing and automation applications to link and manage devices in a network. It was created by Allen-Bradley, a company that is now a part of Rockwell Automation, and it belongs to the CIP family of industrial protocols. DeviceNet is made for easier data communication and to control information for industrial devices including sensors, actuators, motor drives, and programmable logic controllers (PLCs). Here are some essential DeviceNet characteristics and information:

Physical Layer: Multiple devices can be connected in series along a single communication line thanks to DeviceNet's two-wire, multi-drop RS-485 communication network. Depending on the needs of the network, it can also handle larger cable (for longer distances) or thinner cable (for shorter distances).

Topology: The topology used by DeviceNet is typically linear or trunk line-drop. All devices are connected through a trunk line, and individual devices are connected to it by drops (also known as stubs). Additionally, it is capable of supporting a hybrid architecture in which many trunk lines are interconnected.

Data Transmission: A master device (typically a PLC) manages and communicates with numerous slave devices (such as sensors and actuators) using the master-slave communication architecture that DeviceNet employs. It allows acyclic data transfer for configuration, diagnostics, and other non-time-critical data and cyclic data transfer for real-time control applications.

Speed: Depending on the implementation and network configuration, DeviceNet typically runs at data transmission rates of 125 or 500 kbps.

Addressing: Each DeviceNet device on the network is given a distinct address, enabling communication between the master device and particular devices.

Object-Based Communication: Data is structured into objects in the object-oriented communication model used by DeviceNet. It is a versatile and scalable protocol because devices communicate by reading and writing data items.

Integration: DeviceNet is frequently used in conjunction with other products from Rockwell Automation and is thoroughly integrated into their automation systems.

Certification: Devices must pass certification exams given by groups like the ODVA (Open DeviceNet Vendor Association) in order to ensure compatibility and adherence to the DeviceNet standard.

Applications: DeviceNet is frequently used for operations including monitoring and controlling sensors, motor drives, and other industrial equipment in a variety of industries, including the production of automobiles, food and beverage, packaging, and material handling.

ETHERNET/IP

A popular industrial communication protocol used in industrial automation and control systems is Ethernet/IP (Ethernet Industrial Protocol). It is an expansion of the Ethernet standard protocol that satisfies the particular requirements of industrial applications. It is a member of the CIP family of protocols. High data transfer rates, flexibility, and seamless integration into Ethernet-based networks are all hallmarks of Ethernet/IP. Here are some essential Ethernet/IP characteristics and information.

Ethernet-Based Communication: It is simple to integrate Ethernet/IP into existing Ethernet networks since it employs common Ethernet hardware and infrastructure, like Ethernet cables and switches. It communicates via IP-based technologies by utilizing the TCP/IP and UDP/IP protocols.

Real-Time and Non-Real-Time Communication: A variety of industrial automation applications can be handled by Ethernet/IP since it allows both real-time and non-real-time communication. Cyclic I/O (Input/Output) for deterministic control is one method for real-time communication.

Object-Oriented Model: Ethernet/IP utilizes an object-oriented communication model, just like other CIP protocols. Reading and writing data items during device communication enables flexibility and simple configuration.

Scalability: Ethernet/IP is extremely scalable and supports both small-scale industrial installations with a handful of devices and large-scale deployments with hundreds or thousands of units. PLCs,

I/O modules, drives, and a variety of sensors and actuators are among the devices that it supports.

Data Transfer Rates: Depending on the network architecture and device capabilities, Ethernet/IP can achieve data transmission rates of 10 Mbps (Ethernet), 100 Mbps (Fast Ethernet), or even 1 Gbps (Gigabit Ethernet).

Integration with IT Networks: Ethernet/IP networks can be linked with IT networks, enabling seamless data interchange between enterprise-level systems like MES (Manufacturing Execution Systems) and ERP (Enterprise Resource Planning) systems and the factory floor.

Vendor-Neutral: Because Ethernet/IP is an open standard and is supported by many industrial automation equipment manufacturers, devices from various suppliers can communicate with one another.

Safety and Security: To comply with safety regulations, Ethernet/IP offers ways to integrate safety features such as safety I/O. To secure network connections, security measures like authentication and encryption can be used [2, 14, 16, 23, 25].

Diagnostics and Maintenance: Ethernet/IP provides diagnostic tools for analyzing network problems and keeping track of device condition. Maintenance procedures are made easier by the support for remote device configuration and monitoring.

Applications: Various industries, including those that manufacture automobiles, produce food and beverages, manufacture pharmaceuticals, and more, employ Ethernet/IP for operations including machine control, process management, and data collecting.

PROFINET

An industrial communication protocol and network technology called PROFINET (Process Field Network) is employed in the manufacturing and automation sectors. It is intended for real-time data interchange and communication between a variety of industrial devices and systems, such as PLCs, sensors, actuators, human-machine interfaces (HMIs), and other automation parts. High-speed connectivity, adaptability, and scalability are attributes of PROFINET. Key characteristics and information about PROFINET are shown in Table 4.1.

Conformance Classes: Different conformance classes are defined by PROFINET to meet the needs of various applications, from simple automation jobs to high-performance motion control and safety-critical applications.

Scalability: Due to PROFINET's excellent scalability, it can be applied to both small- and large-scale industrial installations that contain thousands of units. Devices including common PLCs, motion controllers, drives, remote I/O modules, and safety equipment are supported.

Table 4.1 Characteristics and information about PROFINET

Ethernet-based communication	Since PROFINET is based on Ethernet standard technology, Ethernet gear and infrastructure are compatible with it. It communicates using the TCP/IP and UDP/IP protocols, which enables simple integration with IT networks.
Real-time communication	Both real-time and non-real-time communication are supported by PROFINET. Mechanisms like PROFINET IO, which guarantee predictable data sharing for time-critical applications, enable real-time communication.

Integrated Diagnostics: Real-time network and device health monitoring is made possible by PROFINET's powerful diagnostic features. Detailed error information can assist with maintenance and troubleshooting.

Safety Integration: The communication protocol for PROFINET can incorporate safety functionality (PROFIsafe), allowing the execution of safety-related operations over the same network.

Redundancy and Reliability: As a result of PROFINET's network redundancy capabilities, vital applications can continue to run without interruption.

Vendor Neutrality: Since PROFINET is an open standard, a wide range of industrial automation equipment producers support it, ensuring compatibility between products from various suppliers.

Security Features: To shield network communications from illegal access and online dangers, PROFINET provides security measures including authentication and encryption.

IT Integration: The data exchange between the factory floor and higher-level systems like MES and ERP systems is facilitated by the integration of PROFINET networks with enterprise-level IT systems.

Applications: PROFINET is utilized for operations including machine control, production monitoring, and remote diagnostics in a variety of industries, including automotive, manufacturing, packaging, and process control.

CAN (CONTROLLER AREA NETWORK)

Most commonly utilized in the automotive and industrial industries, CAN is a communication protocol and network technology. To answer the demand for a strong and dependable communication system in automobiles, Bosch initially developed it in the 1980s. Since then, it has discovered uses outside the automotive industry in many different sectors. Here are some salient CAN characteristics and information.

Differential Serial Communication: Two wires are used for both data transmission and reception in CAN, which employs a two-wire, differential serial communication technique. Due to the inherent noise protection that this differential signaling provides, CAN is very dependable in loud industrial and automotive applications.

Message-Based Protocol: Devices on the network send and receive messages, commonly referred to as frames, through CAN's message-based protocol. An identification (ID) that is part of every communication establishes its priority. Higher priority communications are represented by lower ID values.

Multi-master Network: Because CAN networks are multi-master, many networked devices can transmit messages to one another in a decentralized manner. Bit-wise arbitration is a technique used by devices to access the network in a non-destructive and deterministic way that prevents data conflicts.

Real-Time Communication: Due to its predictable behavior and minimal latency, CAN is a good choice for real-time communication. It is frequently employed in situations where timing and synchronization are essential, like in car engine control systems.

Scalability: The number of nodes (devices) in CAN networks can be simply increased or decreased. Simple two-node networks and complicated networks with dozens of devices are both possible with CAN-based systems.

Data Rates: The most popular data speeds that CAN supports are CAN 2.0A (standard) and CAN 2.0B (high speed). The maximum data rate for CAN 2.0A is 125 kbps, while the maximum data rate for CAN 2.0B is 1 Mbps.

Error Detection and Handling: Devices can detect faults including data corruption and message collisions thanks to the robust error detection features provided by CAN. Retransmission of messages and error frames are two examples of error-handling methods.

Message Prioritization: CAN communications are ranked in priority according to their IDs. If the bus is crowded, messages with lower ID values have a higher priority and may preempt ones with higher IDs.

Applications: Engine management, transmission control, vehicle diagnostics (OBD-II), and in-car networking are just a few of the uses for CAN in the automobile sector. CAN is used in industrial automation for activities like sensor networks, process automation, and machine control.

Variants: In addition to J1939, which is used in heavy-duty vehicles, and DeviceNet, which is used in industrial automation, CAN includes a number of versions. CANopen is one of these.

FOUNDATION FIELDBUS

In sectors like oil and gas, chemical production, and pharmaceuticals, process control and automation applications use FOUNDATION Fieldbus, a digital communication protocol and network technology. It gives different field devices, including sensors, actuators, and controllers, a consistent method for communicating and exchanging data in real time within a process control system. Here are some salient characteristics and specifics of FOUNDATION Fieldbus.

Digital Communication: Given that FOUNDATION Fieldbus is a fully digital communication protocol, it can convey both control commands and process data by using digital signaling.

Multi-drop Topology: In a multi-drop topology, which involves sharing the same communication wire pair across several field devices linked in parallel, FOUNDATION Fieldbus is frequently used. Comparing this design to conventional analog 4–20 mA systems will reduce wiring complexity and expenses.

Time-Sensitive Networking (TSN): TSN technology is used in some FOUNDATION Fieldbus implementations, which improves the protocol's real-time capabilities and makes it appropriate for applications with strict timing constraints.

Data Objects: Data is arranged into objects in FOUNDATION Fieldbus, each of which has a unique set of properties and methods. By reading and writing data items, devices may easily configure themselves and communicate with one other.

Integrated Diagnostics: FOUNDATION Fieldbus has built-in diagnostic tools that aid in locating network failures, device health, and other problems. These diagnostic tools can help with troubleshooting and preventive maintenance.

Segmentation: Large FOUNDATION Fieldbus networks can be segmented into multiple segments, each with its own power source and terminator. Segmentation helps prevent network failure in case of a fault or device failure.

Interoperability: FOUNDATION Fieldbus is built on global standards (such as IEC 61158 and IEC 61784), guaranteeing interoperability between devices from various manufacturers.

Safety Integration: Safety instrumented functions (SIFs) can be used by FOUNDATION Fieldbus to incorporate safety functionality in order to meet safety-critical criteria.

Data Rates: Depending on the exact implementation and network setup, FOUNDATION Fieldbus runs at data speeds ranging from 31.25 kbps to 1 Mbps.

Applications: FOUNDATION Fieldbus is utilized in sectors including chemical manufacturing, oil and gas refining, and power generation where precise process control and data collecting are crucial. It is frequently used in industrial operations to monitor and control things such as temperature, pressure, flow, and level.

HART (HIGHWAY ADDRESSABLE REMOTE TRANSDUCER)

A popular industrial communication protocol called HART (Highway Addressable Remote Transducer) enables two-way communication with field devices in process control and automation systems. HART is particularly used in sectors like manufacturing, chemical processing, and oil and gas, where precise field instrument control and monitoring are crucial. Here are some essential HART characteristics and information:

Hybrid Communication: To mix analog and digital communication on a single two-wire loop, HART is a hybrid communication protocol. The principal process variable is transmitted using the analog signal (4–20 mA current loop), and supplementary data, device diagnostics, and setup are transmitted using the digital signal.

Two-Wire Loop: In contrast to conventional discrete wiring for each device, HART operates via a two-wire communication loop, which simplifies installation and cabling. By adding a HART-compatible modem or transmitter, existing 4–20 mA current loops can frequently be readily updated to allow HART communication.

Master-Slave Communication: A master-slave communication model is used by HART. The field devices, which function as slaves and provide the needed information, receive requests from the master device, which is frequently a distributed control system (DCS) or a mobile communicator.

Two Communication Modes: Point-to-Point (P2P) and Multi-Drop (MD) are the two communication methods that HART supports. Direct connection between a single master and a single field device is accomplished via P2P mode. Multiple field devices may be connected in parallel on the same loop while using MD mode, and the master will successively communicate with each device.

Dual Communication Channels: HART devices feature two communication channels: the digital HART channel for bi-directional communication, and the analog 4–20 mA channel for continuous data transmission.

Device Parameterization and Configuration: Field equipment can be remotely configured and parameterized via HART. Without physically contacting the device, operators can set calibration values and adjust device settings.

Device Diagnostics: With the aid of HART devices, predictive maintenance and troubleshooting are made possible by the vital diagnostic data they provide, including device health, status, and alerts.

Data Types: Process variables (such as pressure and temperature), device status, device warnings, and configuration data are just a few of the types of data that HART can communicate.

Data Rates: HART transmits setup and diagnostic data at a speed of 1,200 bps (bits per second), which is adequate for most applications.

Applications: HART is frequently used to monitor and manage a variety of field devices, such as flow meters, level sensors, pressure transmitters, and temperature sensors. Particularly beneficial are applications that require precise control and diagnostics, such as those found in the process industries.

MODBUS TCP/IP

A popular industrial communication protocol called MODBUS TCP/IP expands the original MODBUS protocol to make it work with Ethernet and TCP/IP networks. It is frequently used to communicate with numerous devices, including PLCs, sensors, actuators, and other equipment in industrial automation and control systems. Here are some essential MODBUS TCP/IP characteristics and information:

Ethernet-Based Communication: Ethernet and TCP/IP are the communication protocols used by MODBUS TCP/IP. As a result, it can use network infrastructure and technology that are common for Ethernet. Typically, it utilizes Ethernet networks, both wired and wireless.

Client-Server Communication Model: A supervisory computer and an HMI are examples of client devices, and a PLC or sensor is an example of a server device in the client-server communication model used by MODBUS TCP/IP. Requests are sent from the client to the server, which answers with data or carries out the desired actions.

Register-Based Communication: A register-based data model is used by MODBUS TCP/IP, where data is arranged into holding registers, input registers, coils, and discrete inputs. Clients can access and modify the data in these registers, which enables the management and observation of field equipment.

Message Structure: A MODBUS Application Protocol (MBAP) header and a MODBUS data payload make up a MODBUS TCP/IP message. Information like the transaction ID, protocol ID, length, and unit ID are all contained in the MBAP header.

Data Types: Various data types, including 16-bit and 32-bit integers, floating-point numbers, and binary data, are supported by MODBUS TCP/IP. These data types are read and written using various function codes.

Data Rates: The Ethernet hardware and network configuration affect the data rates in MODBUS TCP/IP networks. 10 Mbps (Fast Ethernet) and 100 Mbps (Gigabit Ethernet) are typical Ethernet speeds.

Open Standard: Many industrial automation equipment manufacturers accept the open standard MODBUS TCP/IP, providing interchange between devices from various suppliers.

Applications: Many industries, including manufacturing, energy, building automation, and water treatment, employ MODBUS TCP/IP extensively for activities like process control, data collection, and supervisory control. In SCADA systems, it is frequently used to connect devices.

Security Considerations: Given that MODBUS TCP/IP uses TCP/IP networks, implementations should take network security precautions. For communication to be protected, proper access control and firewall configurations are crucial.

Real-Time Considerations: Although many industrial applications can be used with MODBUS TCP/IP, it might not be the best option for those that have strong real-time needs because the communication latency can vary based on network conditions.

OPC (OLE FOR PROCESS CONTROL)

A set of industrial standards and specifications known as OPC, or OLE for Process Control, is used in the fields of industrial automation and process control to facilitate data interchange and communication between various software programs and hardware components. Many industrial equipment, including PLCs, SCADA systems, HMIs, and others, may now communicate with one another seamlessly, thanks to OPC technology. These OPC characteristics and information are essential.

OLE and DCOM: OPC is built on the Distributed Component Object Model (DCOM) and Component Object Model (COM) technologies from Microsoft, although this feature of its initial moniker, OLE (Object Linking and Embedding), has become less significant over time.

Interoperability: OPC is intended to overcome the issue of interoperability in industrial automation systems between hardware and software from various suppliers. Using a standardized interface, it enables software programs to communicate with a variety of industrial equipment and systems.

Client-Server Architecture: Client-server architecture is used by OPC. While the OPC server is in charge of facilitating access to data from multiple devices, the OPC client is an application that makes requests for data or issues commands. OPC servers can be tailored to particular device types or data sources (such as PLCs, DCS, instruments, etc.).

Data Access Standards: OPC covers a number of standards, with OPC Data Access (OPC DA) being the most used. Clients can read and write real-time data from devices via OPC DA. OPC Historical Data Access (OPC HDA) and OPC Alarms and Events (OPC AE) are further OPC standards that provide for the management of real-time alarms and events as well as historical data access.

Vendor-Neutral: OPC supports hardware and software from different manufacturers without the need for special drivers or interfaces because it is vendor-neutral. For their products, vendors generally offer OPC servers, making it simpler to connect them to third-party software.

Security and Authentication: OPC implementations can include security elements to safeguard the integrity and confidentiality of communication, like user authentication and data encryption.

Platform Independence: OPC can be used with a variety of operating systems, including Windows, Linux, and others, and is platform-independent.

Real-Time Data Exchange: OPC makes it possible to transmit real-time data, which makes it appropriate for monitoring and managing industrial processes in industries such as manufacturing, energy, water treatment, and more.

Scalability: OPC can be applied to both small- and large-scale industrial facilities with hundreds or even thousands of units.

Evolution: Newer versions of OPCs have been created as the technology has changed over time. The most recent OPC standard, OPC UA (Unified Architecture), is intended to offer enhanced security, scalability, and cross-platform compatibility.

ETHERCAT

An industrial Ethernet-based fieldbus system and communication protocol called EtherCAT (Ethernet for Control Automation Technology) is frequently utilized in real-time control and automation applications. It was created by Beckhoff Automation and is widely used in many different sectors, including manufacturing, robotics, machine automation, and motion control. Here are some essential traits and information about EtherCAT.

Ethernet-Based Communication: EtherCAT is compatible with current Ethernet technologies since it is built on industry-standard Ethernet hardware and infrastructure. It makes use of the Ethernet frames format and the common Ethernet physical layer, such as Ethernet cables and switches.

Real-Time Communication: EtherCAT is a real-time communication protocol that is renowned for its fast data transfer rates and minimal communication latency. Time-sensitive control applications can benefit from the ability of devices connected to an EtherCAT network to communicate with sub-microsecond synchronization precision.

Distributed Clocks: Distributed clocks are used by EtherCAT to synchronize the network nodes, ensuring accurate device timing. This distributed clock synchronization allows for finely synchronized data transfer between devices as well as coordinated motion control.

Daisy-Chain Topology: Each device in an EtherCAT network is connected to the next in a continuous loop using a daisy-chain or line topology. This topology lowers the complexity of the cabling and offers economical and effective network designs.

Master-Slave Communication Model: A master-slave communication mechanism is used by EtherCAT. Multiple slave devices, including sensors, drives, and I/O modules, are coordinated in communication by a master device, frequently a PLC or motion controller.

Data Processing on the Fly: Each EtherCAT slave device reads and processes data that is relevant to it as data packets flow over the network thanks to the "processing on the fly" method used by EtherCAT. This function lessens the load on the master device and lowers network latency.

Flexible Data Types: Digital and analog signals, motion control data, and process data are all supported by EtherCAT. It can manage the intricate data structures needed for sophisticated control systems.

Hot-Plugging and Auto-Configuration: With the help of EtherCAT, devices can be added to or deleted from the network without necessitating a network restart. Features for automatic configuration make it easier to install and commission new devices.

Safety Features: By ensuring secure device connection, EtherCAT (FSoE) is an extension that makes safety-critical applications possible.

Applications: Applications for EtherCAT include factory automation, robotics, machine tools, packing machinery, and auto assembly lines. It works well for jobs that call for synchronized motion, deterministic control, and high speed.

CANOPEN

Based on the CAN protocol, CANopen is a higher-layer communication protocol and device profile definition. It is frequently utilized in embedded control and industrial automation systems to promote machine [10, 15] and device interoperability and communication. Devices from various manufacturers can communicate smoothly over a CAN network because of the common protocols, object dictionaries, and device profiles defined by CANopen. Here are several CANopen salient characteristics and specifics.

CAN-Based Communication: The CAN protocol, which offers a stable and deterministic communication platform, is the foundation upon which CANopen is based. To standardize device communication, CANopen expands CAN by specifying certain communication guidelines, message formats, and object dictionaries.

Open Standard: The CiA (CAN in Automation) organization maintains the specifications of CANopen, an open standard. This openness encourages compatibility and interoperability between devices made by multiple vendors.

Object-Oriented Communication: As an object-oriented protocol, CANopen divides data into objects, each of which has a specific identity and data type. Since devices communicate by reading and writing objects, setup is flexible and simple.

Device Profiles: Device profiles that define communication parameters, object dictionary entries, and behavior for particular device types are defined by CANopen. Drive, I/O, sensor, and other device profiles are frequently used.

Master-Slave Communication: A master-slave communication architecture underlies CANopen. The slave devices respond to requests or transmit data as necessary when the master device (typically a PLC or controller) establishes communication with them.

Device Configuration: Devices on the network can be remotely configured and parameterized thanks to CANopen. Applications utilizing dispersed and remote equipment might benefit from the flexibility of being able to change parameters and settings without physically contacting the devices.

Safety Integration: CANopen Safety expands CANopen to incorporate safety-critical tasks, enabling it to comply with safety standards like IEC 61508 and ISO 13849. It is also known as CANopen Safety over CANopen or CANopen Safety over EtherCAT.

Device State Management: To manage devices and handle errors, CANopen defines specific device states, such as pre-operational, operational, and stopped.

Data Types: Numerous data kinds, including integers, floating-point numbers, arrays, and others, are supported by CANopen. Various entries in the object dictionary relate to various kinds of data.

Applications: Applications for CANopen are numerous and include robotics, medical equipment, automobile systems, and factory automation. Applications requiring synchronization, interoperability, and real-time control of devices are particularly well suited for it.

MODBUS RTU

A popular communication protocol in systems for industrial automation and process management is called MODBUS RTU. It is a serial communication protocol that transmits data using the RS-232 or RS-485 standards. The advantages of MODBUS RTU are its simplicity, dependability, and ease of use. Here are some important MODBUS RTU characteristics and information.

Serial Communication: A serial communication protocol known as MODBUS RTU is frequently used with RS-232 or RS-485 interfaces. Due to its suitability for industrial situations, longer communication distances, and capability for multi-point networks, RS-485 is more frequently utilized.

Master-Slave Architecture: Using a master-slave communication mechanism, MODBUS RTU communicates. One or more slave devices reply to requests sent by a master device, which starts communication. Slave devices include sensors, actuators, or other types of control devices, while master devices are frequently PLCs, HMI panels, or software programs.

Frame Format: A frame that houses the start character, address, function code, data, error checking, and stop bits makes up a MODBUS RTU message. The frame format ensures error detection and dependable data delivery.

Modbus Functions: The standard function codes that are supported by MODBUS RTU describe a variety of operations, including reading and writing registers, controlling devices, and obtaining diagnostic data. Reading and writing holding registers, input registers, coils, and discrete inputs are examples of common function codes.

Binary Data: In applications where binary states, analog values, or discrete signals must be transferred, MODBUS RTU conveys binary data and is especially well suited for those types of tasks.

Speed and Data Rate: The serial interface being utilized (RS-232 or RS-485) and the particular implementation determine the data rate of MODBUS RTU communication. Typical baud rates are 9,600, 19,200, and 38,400 bits per second (bps).

Error Detection and Correction: A straightforward CRC is used by MODBUS RTU to identify communication issues. The CRC ensures data integrity and aids in the detection of noise- or other-related mistakes.

Frame Length: Depending on the function code and the quantity of data being transferred, the frame length in MODBUS RTU can change. For transmission, longer frames are split up into numerous packets.

Applications: Several industries, including manufacturing, energy, and building automation, employ MODBUS RTU. It is frequently used for keeping an eye on and managing equipment including sensors, motor drives, pumps, and valves.

Scalability and Simplicity: Since MODBUS RTU networks are scalable, additional devices can be added without requiring extensive network modification. The protocol is a popular option for many applications due to its ease of use and widespread industry recognition.

CC-LINK

Automation and control systems use the high-speed, deterministic industrial communication network technology known as CC-Link (Control and Communication Link). It was created by the Mitsubishi Electric Corporation and has found widespread use across a number of sectors, including manufacturing, transportation, and robots. Strongness, real-time capabilities, and adaptability are hallmarks of CC-Link. CC-Link's salient qualities and specifics are listed below.

Open Network Standard: Since CC-Link is an open network standard, numerous manufacturers are free to create and sell products that are CC-Link compliant. This transparency encourages interoperability and permits creative industrial automation system design.

High-Speed Communication: Real-time control applications can benefit from CC-Link's high-speed, low-latency communication capabilities. Depending on the individual CC-Link model, it can achieve data rates of up to 10 Mbps or higher.

Deterministic Performance: Deterministic performance is a feature of CC-Link's design, guaranteeing that communication and control operations go reliably and with little variance in timing. For situations where accurate synchronization and timing are essential, this is essential.

Flexible Topologies: Network topologies supported by CC-Link include ring, line, and star configurations. Additionally, it permits mixed topologies, allowing for customized network designs to satisfy certain application requirements.

Device Types: PLCs, remote I/O modules, drives, robots, and HMIs are just a few of the many devices that can be connected using CC-Link.

Safety and Motion Control: Through CC-Link Safety, CC-Link provides integrated safety functionality that enables the use of safety-critical applications. For applications demanding precise motion and positioning control, it also features extensive motion control capabilities.

Long Communication Distances: Long-distance coverage provided by CC-Link networks makes them appropriate for use in big manufacturing facilities. Further extending the range of communication is possible with fiber optic lines.

Remote Diagnostics and Maintenance: To remotely monitor and troubleshoot network and device issues, CC-Link offers diagnostic functions. This makes preventive maintenance easier and cuts down on downtime.

Integration with IT Networks: The factory floor and more advanced enterprise systems, such as MES and ERP systems, can interchange data thanks to CC-Link's ability to be linked with IT networks.

Global Adoption: CC-Link is utilized all across the world and is widely used in Asia, especially in Japan, where it was first created.

Variants: CC-Link comes in a variety of forms, such as CC-Link IE (Industrial Ethernet), which is based on Ethernet technology and offers better data speeds, and CC-Link Safety, which is targeted at applications that are crucial for safety.

BACNET

Building Automation and Control Networks, or BACnet, is a widely adopted open communication standard and protocol created especially for building automation and control systems (BACS). The American Society of Heating, Refrigerating and Air Conditioning Engineers established it as an international standard, and ISO (International Organization for Standardization) has approved it as ISO 16484-5. The main purpose of BACnet is to make it easier for different BAS and devices to communicate with one another, including HVAC (Heating, Ventilation, and Air-Conditioning), lighting management, access control, fire and life safety, and more. Here are some essential BACnet characteristics and information.

Interoperability: The main objective of BACnet is to guarantee interoperability between BAS and equipment from various vendors. It outlines a standard set of guidelines, data formats, and communication protocols that allow devices from different suppliers to operate together without any issues.

Client-Server Architecture: Client-server communication is the mode of operation for BACnet. Devices connected to a BACnet network can function as clients, requesting data or services, and servers, providing the requested data or services. Sensors, controllers, and supervisory systems can act as clients, whereas access control panels, HVAC controllers, and controllers for lighting can serve as servers.

Standard Object Types: Analog input, binary output, device, and schedule are a few examples of the standardized object types that are defined by BACnet and reflect typical building automation operations. These object types are implemented by devices, making it simpler for integrators and users to comprehend and configure them.

Data Types and Services: Data types supported by BACnet include character strings, floating-point numbers, and integers. It offers a collection of common services that devices can utilize to communicate with one another, like "Read Property," "Write Property," "Who-Is," and "I-Am."

Network Variants: Ethernet, BACnet/IP (over Internet Protocol), BACnet MSTP (Master-Slave/Token-Passing), and BACnet PTP (Point-to-Point) are just a few of the numerous network protocols that BACnet can operate over.

Security Features: To safeguard communication and data integrity, BACnet provides security features like authentication, encryption, and access control. These security precautions are necessary to protect building control systems from online threats.

Scalability: A few devices or big, complicated BAS with hundreds or thousands of devices can be easily accommodated on BACnet networks.

Integration with Other Systems: Other building systems, such as fire alarm systems, security systems, and energy management systems, can be interconnected with BACnet. This enables the centralized management and observation of several building operations.

Trend Logging and Alarming: For building control and administration, BACnet includes trend tracking and alarming capabilities that allow the collecting of historical data and real-time event notifications.

Global Adoption: The building automation industry regards BACnet as a leading standard that has gained widespread adoption throughout the world.

Real-time protocols for connecting systems, interfaces, and instruments are known as industrial communication protocols. In an industrial network, these protocols guarantee connectivity between systems, equipment, and machines. They give managers more visibility into and control over their business processes. Industrial Automation Protocols integrate manufacturing systems and machine assets to provide visibility into production efficiency and equipment health. By capturing machine data at the edge with a variety of industrial control system (ICS) protocols, manufacturers may automate data collection.

REFERENCES

1. Neumann, P. (2007). Communication in industrial automation – What is going on? *Control Engineering Practice*, 15(11), 1332–1347.
2. Bangemann, T., Karnouskos, S., Camp, R., Carlsson, O., Riedl, M., McLeod, S., ... & Stluka, P. (2014). State of the art in industrial automation. *Industrial Cloud-Based Cyber-Physical Systems: The IMC-AESOP Approach*, 23–47.

3. Pereira, C. E., & Neumann, P. (2009). *Industrial Communication Protocols* (pp. 981–999). Springer Handbook of Automation.
4. Huitsing, P., Chandia, R., Papa, M., & Shenoi, S. (2008). Attack taxonomies for the Modbus protocols. *International Journal of Critical Infrastructure Protection*, 1, 37–44.
5. Morris, T. H., Jones, B. A., Vaughn, R. B., & Dandass, Y. S. (2013, January). Deterministic intrusion detection rules for MODBUS protocols. In *2013 46th Hawaii International Conference on System Sciences* (pp. 1773–1781). IEEE.
6. Fovino, I. N., Carcano, A., Masera, M., & Trombetta, A. (2009, March 23–25). Design and implementation of a secure Modbus protocol. In *Critical Infrastructure Protection III: Third Annual IFIP WG 11.10 International Conference on Critical Infrastructure Protection*, Hanover, NH, Revised Selected Papers 3 (pp. 83–96). Berlin Heidelberg: Springer.
7. Willig, A. (2003). Polling-based MAC protocols for improving real-time performance in a wireless PROFIBUS. *IEEE Transactions on Industrial Electronics*, 50(4), 806–817.
8. Drahoš, P., & Bélai, I. (2012). The PROFIBUS protocol observation. *IFAC Proceedings*, 45(11), 258–263.
9. Schiffer, V. (2001, October). The CIP family of fieldbus protocols and its newest member-Ethernet/IP. In *ETFA 2001. 8th International Conference on Emerging Technologies and Factory Automation. Proceedings (Cat. No. 01TH8597)* (pp. 377–384). IEEE.
10. Silva, M., Pereira, F., Soares, F., Leão, C. P., Machado, J., & Carvalho, V. (2015). An overview of industrial communication networks. *New Trends in Mechanism and Machine Science: From Fundamentals to Industrial Applications*, 933–940.
11. Brooks, P. (2001, October). Ethernet/IP-industrial protocol. In *ETFA 2001. 8th International Conference on Emerging Technologies and Factory Automation. Proceedings (Cat. No. 01TH8597)* (Vol. 2, pp. 505–514). IEEE.
12. Brooks, P. (2001, October). Ethernet/IP-industrial protocol. In *ETFA 2001. 8th International Conference on Emerging Technologies and Factory Automation. Proceedings (Cat. No. 01TH8597)* (Vol. 2, pp. 505–514). IEEE.
13. Lindner, S., Häberle, M., & Menth, M. (2023). P4TG: 1 Tb/s traffic generation for Ethernet/IP networks. *IEEE Access*, 11, 17525–17535.
14. Walz, A., Niemann, K. H., Göppert, J., Fischer, K., Merklin, S., Ziegler, D., & Sikora, A. (2023, July). PROFINET security: A look on selected concepts for secure communication in the automation domain. In *2023 IEEE 21st International Conference on Industrial Informatics (INDIN)* (pp. 1–6). IEEE.
15. Tapia, E., Sastoque-Pinilla, L., Lopez-Novoa, U., Bediaga, I., & López de Lacalle, N. (2023). Assessing industrial communication protocols to bridge the gap between machine tools and software monitoring. *Sensors*, 23(12), 5694.
16. Gerodimos, A., Maglaras, L., Ferrag, M. A., Ayres, N., & Kantzavelou, I. (2023). IoT: Communication protocols and security threats. *Internet of Things and Cyber-Physical Systems*.

17. Pereira, C. E., Diedrich, C., & Neumann, P. (2023). Communication protocols for automation. In Springer *Handbook of Automation* (pp. 535–560). Cham: Springer International Publishing.
18. Kostadinovic, M., Dobrilovic, D., Jotanovic, G., Jausevac, G., Stojanov, Z., & Brtka, V. (2022, June). Analyzing performance of wireless network based on the industrial HART protocol. In *6th EAI International Conference on Management of Manufacturing Systems* (pp. 99–113). Cham: Springer International Publishing.
19. Rahman, A., Mustafa, G., Khan, A. Q., Abid, M., & Durad, M. H. (2022). Launch of denial of service attacks on the Modbus/TCP protocol and development of its protection mechanisms. *International Journal of Critical Infrastructure Protection*, 39, 100568.
20. Yu, F. (2022, December). Research on the principle and implementation of network attack in industrial control environment. In *2022 IEEE 5th Advanced Information Management, Communicates, Electronic and Automation Control Conference (IMCEC)* (Vol. 5, pp. 89–92). IEEE.
21. Kim, D., Lee, J., Lee, S. H., & Kim, S. (2023, May). High-bandwidth control structure for solid-state-transformers with EtherCAT protocol. In *2023 11th International Conference on Power Electronics and ECCE Asia (ICPE 2023-ECCE Asia)* (pp. 1843–1848). IEEE.
22. Kekre, A. M., & Kothari, A. (2022). MODBUS-TR: Advanced MODBUS-RTU protocol for IoT with auto-discovery and triggers. *Wireless Personal Communications*, 125(3), 2769–2780.
23. Yu, G., & Shon, T. (2022, August). Security and forensic analysis for industrial ethernet protocols. In *2022 International Conference on Platform Technology and Service (PlatCon)* (pp. 63–65). IEEE.
24. Sankar, P., Vallikannu, R., Justin, G., & Karg, S. (2023, April). Ambient intelligence based LED lighting control system using BACnet protocol. In *2023 International Conference on Artificial Intelligence and Applications (ICAIA) Alliance Technology Conference (ATCON-1)* (pp. 1–5). IEEE.
25. Zeng, M. (2022, October). A review of smart buildings protocol and systems with a consideration of security and energy awareness. In *2022 IEEE 13th International Green and Sustainable Computing Conference (IGSC)* (pp. 1–6). IEEE.

Chapter 5

Market Watch

Anurag Mishra, Nishant Jawla,
Satyam Sharma, and Pranjal Rai

INTRODUCTION

Users can use our virtual currency to purchase stocks and other securities on Market Watch in real time. Users will be able to experiment with various investing strategies without risking their real money as a result. Users will also be able to look at and take notes from the top performers at the same time. Market Watch is a stock prediction program that generates predictions using information from both formal and informal sources. A few examples of official sources include past prices and trends, news, and so forth. However, informal sources such as platforms, tweets, and comments offer emotional elements, which are crucial factors in determining stock values. We develop a platform that monitors user activity, holds contests, offers a virtual trading playground for collecting extra insights, and has comment sections where we may gather more data. We use all of this data to train our model and estimate stock prices. The main objective of the research is to better understand how emotional qualities interact with robust attributes in the prediction process. The overall architecture is shown in Figure 5.1, which is a 2D representation of the project schema. It displays the experimental flow chart or project architecture for our suggested stock price forecasting algorithm.

LITERATURE REVIEW

A. Hogenboom et al. [1] presented their behavioral finance research. With accurate information, predicting the stock market has proven to be extremely difficult. In this study, they forecast stock prices using cutting-edge online data sources. They discovered that individual investors regularly use online social media platforms to learn about equities in several developing countries (like China). Therefore, it is likely that data collected from social media sites will include relevant information. Using information from East Money, China's top social media platform, and the stocks that users were following, they created daily social networks. Then, they utilized the LSTM model

DOI: 10.1201/9781003479031-5

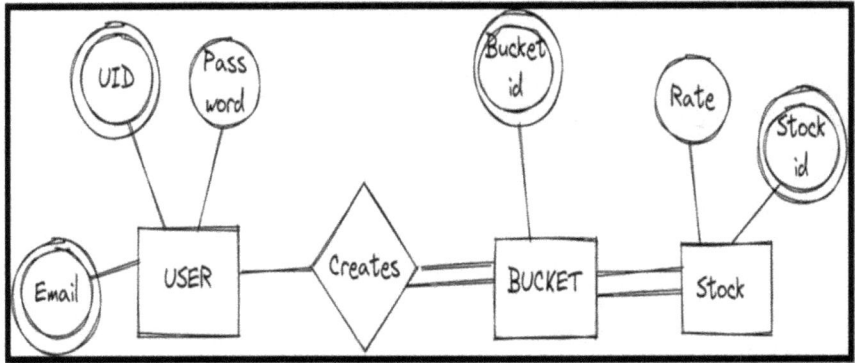

Figure 5.1 Architecture of proposed schema.

to forecast the close prices of the SSE 50 constituent stocks as an addition to conventional variables, calculating the network variable for each stock. Their empirical research demonstrates that including social networks as a variable during prediction might enhance accuracy greatly. These results might help investors predict the future more accurately.

Liu et al. [2] created cutting-edge decision support technologies that were applied to stock price forecasting. They also included text event detection, which was based on trends, in their forecasts. Despite the fact that word sense disambiguation is crucial for text comprehension, these methods typically overlook the meaning of words. They thus proposed an enhanced NLP pipeline for stock price prediction based on events that integrate word sense disambiguation into the detection procedure. They take occurrences from news articles written in plain language and give them weight based on newspaper archives. They assessed word sense disambiguation's benefits for the NASDAQ-100 corporations.

A study on forecasting stock prices is presented by Park and Shin [3]. They discovered that stock prices are impacted by a variety of factors, including economic conditions, foreign exchange rates, interest rates, and oil prices. Although each of these factors has an individual effect on stock prices, the combined effect of their intricate connections or network architecture is stronger. The interplay and complexity of these issues are not accurately reflected by the resource forecasting techniques now in use. This chapter suggests a stock prediction method based on a semi-supervised learning (SSL) algorithm to overcome these drawbacks. With the use of the SSL algorithm, a network may be built, with nodes standing in for elements and edges for similarities between them. Due to network structure, SSL methods may depict the cyclical and reciprocal effects of various elements on predicting.

The work of Long et al. [4] sparks debate regarding the predictability of stock prices. This study suggests a hybrid stock price prediction model that combines deep learning with a sentiment analysis methodology. They classified the investors' concealed sentiments, which were taken from a well-known stock forum, using a Convolutional Neural Network model. They developed a hybrid research model using the results of the first step's sentiment analysis along with the Long Short-Term Memory (LSTM) Neural Network technique to examine technical indicators from the stock market. To verify the efficacy and applicability of the suggested methodology, they conducted real-world experiments on six important industries at the Shanghai Stock Exchange over three time intervals. The experiment's findings demonstrated that the suggested model outperformed baseline classifiers in its ability to categorize investor sentiment and that this hybrid approach outperforms both the single model and models lacking sentiment analysis in its ability to predict stock prices.

Several researchers such as Pahwa et al. [5] have attempted to forecast stock price movement using financial news and machine learning (represented by SVM, or Support Vector Machine). However, the majority of them concentrate on the news content, while only a small number consider the information concealed in the connections between various stories. They suggest a novel kernel based on SVM in this study that takes into account both the contents and the information structures that connect them. The semantic and structural kernel, or S&S kernel, is so called because both the information structures and the news content are imported into its kernel. The efficiency of their kernel is demonstrated using financial news from the medical sector. They discovered that their technique surpasses the others by at least 5%, which is a substantial improvement when comparing the prediction accuracy of the S&S kernel to that of other kernels, such as the linear kernel. The outcome further supported the idea that the informational framework in the daily financial news can provide additional information that aids in predicting the trajectory of stock price.

All of the researchers discussed in Shah [6] present their work on an online trading platform that has already transformed the way consumers buy and sell equities. The world's financial markets have quickly developed into a powerful, linked marketplace. These innovations create many new issues while also opening up new possibilities. Modern stock markets, in contrast to conventional ones, are constructed utilizing a variety of cutting-edge technologies, including machine learning, expert systems, and big data, which interact with one another to promote better decision-making. In addition, with increasing global connectivity to the internet and the prevalence of news, the stock market has become more volatile and susceptible to hostile attacks. How the stock market develops can be significantly predicted by conducting more research in this area.

In a market with many demand segments, the work presented by Genc and De Giovanni [7] examines green investment and pricing options. This document's specific inclusion of ethical customers who examine raw materials while being emotionally and environmentally sensitive is one of its novel aspects. Some emotionally driven consumers base their purchasing decisions solely on researching a company's environmental impact, which is demonstrated by the environmental efforts amassed over time and the emotions linked with them. It's interesting that these folks avoid pricing because of some unfavorable attitudes. On the other hand, a more cautious buyer is primarily motivated by both force and a price reduction plan. The buyer would do well to check all the conveniences before making a raw purchase. In addition, some consumers are as savvy as they are knowledgeable and research pricing before making a purchase. They also look at how each customer affects establishing a sound strategy or plan that will evolve through time, i.e., with green attachments.

Jeong and Jung [8] used a social media-based strategy and undertook a thorough investigation to ascertain whether social status (as represented by internet memes) accurately reflects the emotional content of the stock market. They carried out Granger Causality tests to establish causative linkages between variables, forecasting analysis of the random forest model, and line regression analysis to establish statistical relationships between social variability and stock market prices. The findings demonstrate a statistically significant association between the range of public sentiments associated with the emotional intensity of memes and the user's involvement in memes. Double-r scores in both linear regression analysis and random forest forecasting study show that large snowfall and US-based indicators are highly associated with social mood volatility. When periodically reviewing the database, they observed that there are differences in the degree of social standing represented in the stock market indicators. They come to the conclusion that social status information is fully integrated into both stock market indicators and memes.

With regard to tensions and wars around the world, Tricky and Ben [9] look at the association between the US stock market and the price of gold. They also present an up-to-date indicator called the Geopolitical Risk Index, which was created in 2016 by Caldara and Iacoviello. This study emphasizes the advantages of including gold in your investment portfolio. View the S&P 500, the Geopolitical Risk Index, and the monthly gold price from January 1985 to December 2018. Examine the connection between market volatility and recovery using the MV-GARCH model and the dynamic copular. According to Power's findings, the S&P 500 has a weaker relationship with gold during peaceful times (i.e., a low GPR) than it does amid overt political radicalism. This demonstrates that gold is a good delineator and safe haven, particularly in times of great stress. In addition, we discovered that at times of high volatility, gold has a particularly significant impact in reducing the volatility of the S&P 500.

Research published by Hao et al. [10] reveals emotional information and underlying themes in news items. The new twin vector machine is designed to ingest a substantial amount of data from internet news sources and utilize it to forecast patterns in stock prices. This approach makes extensive use of abstract set theory. This is due to the usage of ambiguous terminology in the text, such as top/bottom and the blurred line between ascending and descending order. The goal of the decision-making process in partial segment order relationships is typically to keep stock prices within a given range and to represent the degree of membership in these stocks favorably (trend). It's helpful to use suggested strategies when conversing with strangers. The dimensions of the illogical hyperplane mode do not change, but outliers only marginally widen the fuzzy solution border. As a result, when compared to other approaches that incorporate outside factors into the data, our model is substantially more reliable. A smaller membership range and a distinct decision margin also make it easier to comprehend the degree of confidence in the outcome. Any application of decision-making must take the evident feature of confidence into consideration.

Research published by Koratamaddi et al. [11] shows that one of the most intricate financial modeling techniques in use today is the stock market. Finding the best investment strategy for assembling a portfolio of chosen stocks that successfully increase returns while reducing related risks is therefore a challenging issue when making the decision to allocate a portfolio of stocks. By training a smart agent at past stock prices, in-depth consolidation learning algorithms have demonstrated promising outcomes when applied to automate portfolio distribution. To better understand and analyze portfolios, however, modern investors are actively engaged in digital forums like social media and online news websites. Market sentiment is the overall judgment that investors have about particular stocks or financial markets. Existing methods do not take into account market sentiment, which has been proven to have a significant impact on investor choices. In our work, we suggest a thorough novel-enhancing learning curve to effectively train a smart automotive trader who not only analyzes historical stock price data but also observes stock market mood by Dow Jones firms. It demonstrates the strength of our strategy when compared to the fundamentals of all conventional measures, such as Sharpe value and ROI.

The work of Otero-Gonzalez and Duran-Santomil [12] combines excellent knowledge with hard data to examine whether the combined investment will outperform that of rivals. When the scores are analyzed separately, the findings provide less support. Mutual funds with robust management, sound practices, and favorable costs enhance the selection process. Last but not least, combined financial estimations are improved by the integration of better (quantitative) and analytical (qualitative) information and its components.

Nikolay Russanov et al. [13] discovered that in the mutual fund industry, around a third of all costs are associated with effective management costs. We discovered that marketing is just as crucial to a fund's size as that of performance and payment when searching for high-end investors and analyzing a fund's potential to assess its construction model. The elimination of marketing greatly enhances well-being. Lower payments result from shifting capital to less expensive media and competition. Budget allocation and operating cost reduction are tightly related. Reducing marketing and administrative expenses while changing to idle investments are also ways to lower the cost of seeking investments.

Ricardo Laborda and coworkers [14] published their research on the effects of the spread of terrorism on the Spanish stock market between 1993 and 2017. Domestic Terrorism (ETA) and International Terrorism Associated with Islamic Extremism are the two components that make up the Daily Terrorist Index, which displays the terrorist activity across various criminal contexts. Because it explains almost half of the variance in the prediction error, their statistical analysis demonstrates the significance of the link, largely as a result of the effects of a terrorist attack. Their dynamic research suggests a higher likelihood of bloodshed during the initial ETA attack period.

Iqbal and Bardwell [15] peg the cost of terrorism to the economy in 2018 at $33 billion. The cost of terrorism to the global economy from 2000 to 2018 was $855 billion. The model adopts the methodology of the 2019 Global Terrorism Index and incorporates the cost of the four indicators from terrorist incidents using cost-effective calculating techniques. Terrorism-related fatalities, injuries, property destruction, and GDP loss are the four indicators. This article's conclusions indicate that international terrorism worsened in 2014, killing 33,555 people and causing economic losses of $111 billion. Terrorism-related deaths rose by 353% from 2011 to 2014, while terrorist attacks rose by 190%. One hundred cases of serious economic loss resulting from injury and death were included in the analysis. Both the $40.6 billion attack on the US on September 11, 2001, and the $4.3 billion Sinjar slaughter in Nineveh Sinjar, Iraq, resulted in significant economic damages.

Kumar, Kandhamal, and Dattatray P. [16] provided a survey of many methods used to make accurate stock market predictions, which were divided into prediction methods and clustering methods. With the aid of 50 research articles, this survey's objective was to categorize the current strategies according to their publication dates, methodology that has been adopted, datasets that have been used, performance measures, and implementation tools. Approaches such as ANN, SVM, SVR, HMM, NN, fuzzy-based techniques, K-means, and others are used to anticipate the stock market. To provide a useful future scope, the research gaps and problems with stock market forecasting are further elaborated. ANN and fuzzy-based techniques are the most frequently used methods for achieving accurate stock market forecasts. These methods can be successfully applied for managing and keeping an eye on the entire stock market. The main issue facing stock price prediction systems is that the majority of the methods currently in use cannot be identified using historical stock data since they are influenced by a variety of factors, including governmental policy decisions, market attitudes, and other considerations. As a result, data from many sources is needed for decision-making, and data pre-processing is a challenging task for data mining.

Moghar, Adil and Hamiche, Mahmed [17] forecast the future value of GOOGLE and NKE assets using an LSTM-based RNN. The results of the model have shown some encouraging signs. The test outcomes demonstrate that the model is capable of following changes in the open prices for both assets.

Anita Yadav and coworkers published their research with Indian stocks as follows. In Experiment 1, LSTMs from ICICI, TCS, Reliance, and Maruti were contrasted with stateful and stateless LSTMs. P-values indicated that there was no statistically significant difference between stateful and stateless LSTMs for the experiment's chosen stock price prediction problem. The random seeding that occurs before each LSTM run, which results in little variances in output, can be used to explain the majority of the value differences. The box and whisker plot diagram's standard deviation values and spread demonstrate that the deviation or spread for stateless is smaller than that for stateful LSTM. As a result, it appears that stateless LSTM is more stable than stateful LSTM. It may be argued that a stateless LSTM model is preferable for time series prediction issues due to higher stability. It makes sense to use stateful LSTM when performing language modeling where subsequent batches are subsequent pieces of text with potential links. The number of concealed layers in Experiment 2 ranged from 1 to 7. The findings indicate that the optimal design for RMSE is $n = 1$. A one-way ANOVA test supported this conclusion as well. Due to improved accuracy, simpler training, and a lower danger of overfitting, we suggest employing one hidden layer to tackle the majority of issues in this situation. More layers may be required if you have extremely complicated problems that cannot be accurately modeled in a single layer, but these are anticipated to be uncommon. The LSTM benefits from increased hidden layer density by becoming more stable. By lowering the box and whisker plot spread and standard deviation values, this is shown to be the case.

PROPOSED METHODOLOGY

Many people, especially students and novices, desire to learn more about the stock market, but they frequently incur significant financial losses. They require a platform where they can observe the market firsthand without the actual risk of losing money. Creation of a secure environment for learning the trades and experimenting with various investment methods involves developing a stock prediction tool using formal and informal data sources, as well as evaluating the impact of emotional elements on stock prediction.

Create a cross-platform stock trading platform that allows users to transact in stocks using virtual money. Learn to propose stocks utilizing information from both formal and informal sources. Find the sentiment analysis algorithm that performs the best to understand the emotional property.

Examine how the psychological perspective influences the impact of a person's emotions on stocks. When gathering or giving stock-related advice, also take into account the tweets from social media and news headlines. The model developed for this project will take into account the investors as well as the reliable data that is already available because they are an essential component of the stock market. In a sense, the stock market is driven by investors. In addition to other sources, we will leverage their analysis and opinions on specific stocks in order to better identify future trends, which will aid our model in producing more precise investing recommendations.

AMRA

AMRA (Autoregressive Integrated Moving Average) group of models represent different common temporal structures found in time series data. With the addition of the integration notion, it is a generalization of the more straightforward autoregressive moving average. The underlying process that produced the observations is presumed to be an ARIMA process when using an ARIMA model for a time series.

Recurrent neural networks with the ability to learn long-term data dependencies are known as LSTMs. This is made possible by the repeating module of the model, which consists of four interconnected layers. Three gates and cell states in the LSTM module allow it to selectively learn, unlearn, or save data for each department. There are very few linear interactions possible in LSTMs because of the cell states that let information flow through blocks unchanged. There is a forget gate in every block that can add or remove information about input, output, and cell state. The sigmoid function is used by the forget gate to decide what data from the cell's previous state should be forgotten. The input gate regulates the flow of data about the state of the cell using the point-by-point multiplication operations "sigmoid" and "tang," respectively. The output element ultimately chooses what data should be sent to the following concealed state.

Linear regression

Originally used in the field of statistics to study the relationship between numerical input and output variables, linear regression is now used in machine learning. This approach combines machine learning with statistics. Learning a linear regression model entails estimating the values of the plot's coefficients based on the data at hand.

SVM

SVM is a supervised machine learning algorithm that can be applied to classification and regression issues. It is mainly applied to classification issues. Each data point is converted into a point in an n-dimensional space by the SVM algorithm (where n is the number of features it possesses). Each function's value corresponds to a certain coordinate. Then, we perform the classification by locating a hyperplane that effectively divides the two classes.

SVR

Originally put forth by Drucker et al., SVR is a supervised learning approach built on Vapnik's support vector theory. By creating a hyperplane and minimizing the difference between predicted and observed values, SVR seeks to lower errors. Hidden Markov Models (HMMs) that are models for sequences compute an output sequence of the same length given an input sequence that is equal to a word. In an HMM model, the probability distribution of a label's nodes and the likelihood that an edge would travel from one node to another are represented as a graph. The input sequences can be combined to calculate the likelihood of a specific label sequence. Flutter is a cross-platform, open-source software development kit created by Google. It enables to produce applications that run on Linux, macOS, Windows, Google Fuchsia, Android, iOS, and the web (all using the same code base). One of the top NO-SQL databases is MongoDB, an open-source document database. It is applied to the development of highly adaptable and open web applications.

Express.js is a back-end framework for Node.js. It is designed for the creation of web applications and APIs and is free and open source. User interfaces for single-page web apps are created using the JS framework React. It is looked after by Meta.

The V8 engine is used by Node.js, a JS runtime environment, to execute JS code outside of a web browser. This gives developers the freedom to use JS for server-side scripting or to build CLI tools.

Tailwind

Tailwind offers basic utility classes that enable the creation of wholly original designs. It is a low-level, highly flexible CSS framework that offers the building blocks required to produce unique designs. High-level programming language called Python is typed and interpreted dynamically. In addition, it offers a number of helpful libraries, which makes it the best choice for development.

Docker

Docker is a platform for creating, distributing, and running programs. By swiftly releasing, testing, and deploying code, it decouples the app from the infrastructure, substantially reducing the time between development and production.

AWS

Amazon Web Services, Inc. (AWS) is a company that offers cloud computing services on demand.

Figma

Often used for building user interfaces, Figma is a vector graphics editor and prototype application. Git is a distributed version control system that is open source and free. Any set of files can be tracked for changes with Git. During the software development lifecycle, it is frequently used to manage code across several branches and to achieve nonlinear workflows.

GitHub

Git is simply hosted on the internet by GitHub. In addition to Git's distributed version control and source code management capabilities, it also offers some of its own features, such as code sharing between teams.

RESULTS

A wireframe is an illustration of web pages or app screens based on content priority, functionalities, and intended behavior. We have shown the wireframe for the project in Figure 5.2. Wireframe screens show how our application will operate. A use case diagram that illustrates the system's scope and key features is shown in Figure 5.3. The interaction between the system and its actors is also shown in the diagram. Use case diagrams only display the use cases and specify what the system performs and how actors interact with it; it doesn't reveal how the systems operate internally. It displays the flow between the stock-filled buckets and our platform.

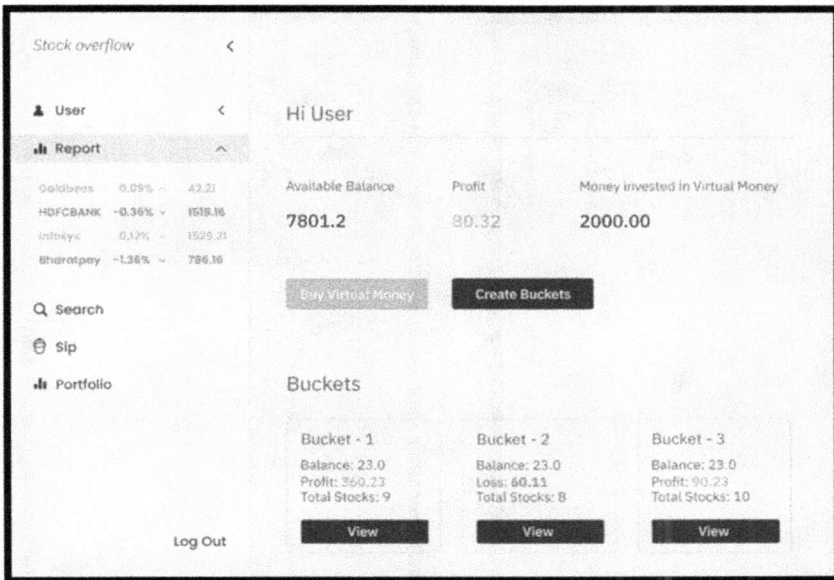

Figure 5.2 Wireframe.

CONCLUSION

Our research into the existing literature has led us to the conclusion that a complete Mobile+Web application will enable users to create portfolios, buy and sell stocks using virtual currency, and experience real-time market activity without actually losing any money. As part of other competitions, we will provide users with a finite amount of virtual currency, and at the conclusion of the competition, we will determine which user has maximized their use of that virtual currency. We'll also create a machine learning (ML) model to forecast probable high-value stocks using sentiment analysis on user-generated data. Then, using graphs and charts, we may educate the users of this information. We might also have some information that is only available to VIP users. We came to understand that investors are a crucial component of the market and that one of the key factors influencing the market is their emotions or biases. As a result, we are able to gather and separate investor opinions and assessments of specific stocks from their market, social media, and other online activity. To predict the market's future trend and develop more successful investment strategies, we can then analyze the feelings and viewpoints that the majority of investors hold.

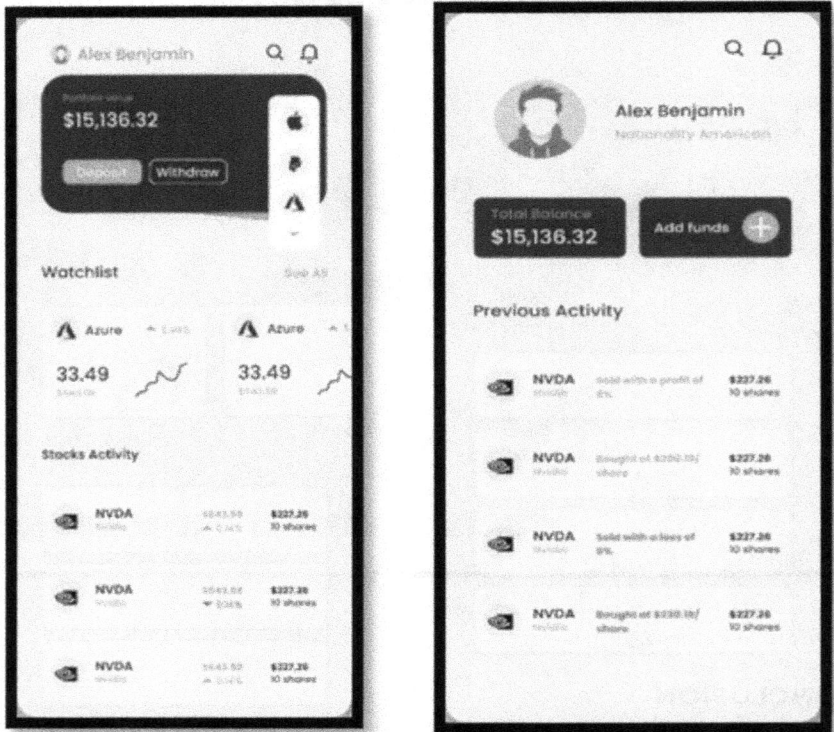

Figure 5.3 Dashboard and outcomes.

REFERENCES

1. Hogeboom, A., Brojba-Micu, A., & Frasincar, F. (2021). The impact of word sense disambiguation on stock price prediction. *Expert Systems with Applications*, 184, 115568. doi:10.1016/j.eswa.2021.115568
2. Liu, K., Zhou, J., & Dong, D. (2021). Improving stock price prediction using the long short-term memory model combined with online social networks. *Journal of Behavioral and Experimental Finance*, 30, 100507. doi:10.1016/j.jbef.2021.100507
3. Park, K., & Shin, H. (2013). Stock price prediction based on a complex inter-relation network of economic factors. *Engineering Applications of Artificial Intelligence*, 26(5–6), 1550–1561. doi:10.1016/j.engappai.2013.01.009
4. Long, W., Song, L., & Tian, Y. (2018). A new graphic kernel method of stock price trend prediction based on financial news semantic and structural similarity. *Expert Systems with Applications*, S095741741830650X. doi:10.1016/j.eswa.2018.10.008

5. Pahwa, K., & Agarwal, N. (2019, February 14–16). *[IEEE 2019 International Conference on Machine Learning, Big Data, Cloud and Parallel Computing (COMITCon)*, Faridabad, India]* – Stock Market Analysis Using Supervised Machine Learning, 197–200. doi:10.1109/COMITCon.2019.8862225

6. Shah, I., & Zulkernine (2019). Stock market analysis: A review and taxonomy of prediction techniques. *International Journal of Financial Studies*, 7(2), 26. doi:10.3390/ijfs7020026

7. Genc, T. S., & De Giovanni, P. (2021). Dynamic pricing and green investments under conscious, emotional, and rational consumers. *Cleaner and Responsible Consumption*, 2, 100007. doi:10.1016/j.clrc.2021.100007

8. Jung, S. H., & Jeong, Y. J. (2021). Examining stock markets and societal mood using Internet memes. *Journal of Behavioral and Experimental Finance*, 32, 100575. doi:10.1016/j.jbef.2021.100575

9. Triki, M., & Ben, M. (2020). The GOLD market as a haven against the stock market uncertainty: Evidence from geopolitical risk. *Resources Policy*, 101872. doi:10.1016/j.resourpol.2020.101872

10. Hao, P., Kung, C., Chang, C., & Ou, J. B. (2020). Predicting stock price trends based on financial news articles and using a novel twin support vector machine with fuzzy hyperplanes. *Applied Soft Computing*, 106806. doi:10.1016/j.asoc.2020.106806

11. Koratamaddi, P., Wadhwani, K., Gupta, M., & Sanjeevi, S. G. (2021). Market sentiment-aware deep reinforcement learning approach for stock portfolio allocation. *Engineering Science and Technology, an International Journal*, 24(4), 848–859. doi:10.1016/j.jestch.2021.01.007

12. Otero-González, L., & Durán-Santomil, P. (2021). Is quantitative and qualitative information relevant for choosing mutual funds? *Journal of Business Research*, 123, 476–488. doi:10.1016/j.jbusres.2020.10.015

13. Roussanov, N., Ruan, H., & Wei, Y. (2020). Marketing mutual funds. *The Review of Financial Studies*, hhaa095. doi:10.1093/rfs/hhaa095

14. Laborda, R., & Olmo, J. (2019). An empirical analysis of terrorism and stock market spillovers: The case of Spain. *Defense and Peace Economics*, 1–19. doi:10.1080/10242694.2019.1617601

15. Bardwell, H., & Iqbal, M. (2020). The economic impact of terrorism from 2000 to 2018. *Peace Economics, Peace Science and Public Policy*, 27(2), 227–261. doi:10.1515/peps-2020-0031.

16. Moghar, A., & Hamiche, M. (2020). Stock market prediction using LSTM recurrent neural network. *Procedia Computer Science*, 170, 1168–1173. doi:10.1016/j.procs.2020.03.049

17. Yadav, A., Jha, C. K., & Sharan, A. (2020). Optimizing LSTM for time series prediction in the Indian stock market. *Procedia Computer Science*, 167, 2091–2100. doi:10.1016/j.procs.2020.03.257

Chapter 6

Study on cybersecurity
Trending challenges, emerging trends, and threats

Rabban Javed, Rashmi Vashisth, and Nidhi Sindhwani

INTRODUCTION

Cybersecurity definition

The term "Cybersecurity" refers to the protection of computer systems, networks, and data from being compromised or accessed unlawfully, to prevent damage or loss [1]. The implementation of various technologies, procedures, and practices is used to ensure the safety of digital information and to deter cyber-attacks [2].

Cybersecurity refers to a collection of procedures, tools, and techniques used to safeguard computer systems, networks, data, and information from unauthorized access, assaults, loss, or theft in the rapidly changing digital environment. Here are a few detailed explanations on cybersecurity.

The foundation of digital trust and resilience is cybersecurity, which will be a key feature of the quickly changing technological world. It includes the use of cutting-edge defense mechanisms, such as intrusion detection systems, access controls, encryption, and authentication, to safeguard networks, interconnected devices, and critical infrastructure from cybercriminals, hackers, and state-sponsored actors looking to gain unauthorized access, steal data, or disrupt digital services.

A set of processes, systems, and structures are employed to protect computer networks and other electronic devices from instances where ownership rights are infringed upon [3].

The following are current issues in cybersecurity.

Ransomware Attacks: Files encrypted by ransomware attacks are demanded to be decrypted in return for money. These attacks are increasing day by day and this can be extremely damaging to both persons and corporations.

Internet of Things (IoT) Security: The IoT is a network of hardware, including computers, mobile phones, and appliances, that is linked to the Internet. The risk of cyber-attacks increases along with the quantity of IoT devices. These gadgets frequently have poor security, which leaves them open to hackers [4–7].

DOI: 10.1201/9781003479031-6

Sophisticated Cyber Threats: Cybercriminals continue to build increasingly complex attack methods by utilizing cutting-edge technology like automation and artificial intelligence. Advanced persistent threats (APTs), social engineering assaults, and ransomware all offer serious hazards to people and organizations, resulting in monetary losses, data breaches, and operational delays.

Vulnerabilities in the Supply Chain: Cyber-attacks have made the global supply chain one of their top targets. Attackers aim to take advantage of weaknesses in the linked ecosystem, focusing on cloud service providers, third-party suppliers, and software and hardware components. Supply chain breaches may lead to the compromise of vital systems, unauthorized access to private information, and the spread of malware to several businesses.

Cloud Security: Increasingly businesses are moving their data and their software to the cloud. Data breaches can greatly affect businesses and their clients; thus, cloud security is a crucial worry.

Addressing these cybersecurity threats will need a thorough and multilayered strategy. Technology improvements, sound governance structures, user education and awareness, public-private sector cooperation, ongoing threat monitoring, and threat adaptation are all part of this process.

Cybercrimes

Cybercrime is the newest and maybe most difficult problem facing the online community. "Any illegal conduct that makes use of computer devices whether an instrument, target or as a means of committing additional crimes" is the definition of cybercrime [8].

Cybercrime continues to pose serious hazards to people, businesses, and society at large. Some of the latest cybercrimes are indicated below.

Credential Stuffing: Cybercriminals utilize credential stuffing strategies because multiple data breaches have revealed login and password combinations. They take advantage of people who reuse passwords across several accounts by utilizing automated programs to test stolen credentials on various websites and online services. Successful login attempts provide hackers access to private data, money accounts, or even business networks.

Deepfake Manipulation: Deepfake technology, which uses AI to manipulate or produce convincing fake text, audio, or video material, offers serious concerns. Deepfakes may be used by cybercriminals for social engineering assaults, media manipulation for propaganda, and persona impersonation. This technology can exacerbate the spread of misinformation and make it difficult to distinguish between real material and stuff that has been altered.

SIM Switching: SIM swapping entails moving a victim's phone number erroneously to a SIM card that is within the attacker's control. Cybercriminals can then access accounts, get beyond two-factor authentication (2FA) security measures, and engage in unauthorized actions like draining Bitcoin wallets or taking control of social media accounts.

IoT Exploitation: The spread of vulnerable IoT devices has given attackers new targets. They use flaws in industrial IoT systems, healthcare equipment, or smart home devices to obtain access without authorization, conduct DDoS assaults, or steal confidential data. IoT botnets frequently employ large-scale attacks by combining the power of hacked devices.

BEC and EAC Attacks: Businesses targeted by impersonating executives or reliable connections through email are known as business email compromise (BEC) and email account compromise (EAC), respectively. These assaults frequently seek to trick employees into starting fraudulent wire transfers, divulging private information, or diverting funds to accounts under the control of the attackers. Such assaults may result in severe monetary losses and harm to one's image.

To reduce the risks connected with these developing cybercrimes, it's crucial to remain up-to-date on the newest cyberthreats and make sure that effective cybersecurity measures are implemented. To defend against the most recent cybercrimes, businesses and people should maintain vigilance, implement solid security practices, frequently upgrade software and systems, and become knowledgeable about developing risks.

Cyber-attack

Cyber-attacks are criminal actions by people or organizations with the goal of taking advantage of security flaws in computer systems, networks, or other digital infrastructure. These assaults may aim to obtain unauthorized access, steal confidential information, disturb business as usual, or even cause harm.

Technology advances, and so do the strategies and methods employed by cybercriminals. To minimize the dangers connected with cyber-attacks, it is essential for individuals and businesses to be educated about the most recent threats, put in place strong security measures, and periodically upgrade systems.

Figure 6.1 provides a view of cyber-attacks in different regions of the world.

Significance of the research

Governments, organizations, and people all share a serious worry about cybersecurity. With the development of mobile devices, cybercriminals now have a larger attack surface thanks to the IoT, making it simpler for them to target businesses and people.

CYBER ATTACK CATEGORIES BY REGION

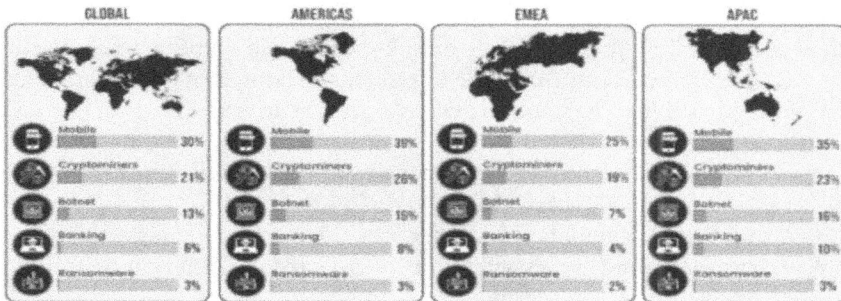

GLOBAL		AMERICAS		EMEA		APAC	
Mobile	30%	Mobile	39%	Mobile	25%	Mobile	35%
Cryptominers	21%	Cryptominers	26%	Cryptominers	19%	Cryptominers	23%
Botnet	13%	Botnet	19%	Botnet	7%	Botnet	16%
Banking	8%	Banking	8%	Banking	4%	Banking	10%
Ransomware	3%	Ransomware	3%	Ransomware	2%	Ransomware	3%

Figure 6.1 Cyber-attack categories by region [9].

The subject of cybersecurity is full of linked discourses. Deconstructing the phrase "cybersecurity" makes it easier to place the conversation within the context of both terms "cyber" and "security" as well as making some of the underlying problems more obvious.

Some significance of the research is as follows:

- Confidential Information Protection
- Continuity of business
- Maintenance of trust
- Compliance

In conclusion, cybersecurity research is crucial for keeping up with emerging threats, creating successful defense plans, and promoting a safe and secure digital environment. It encourages innovation, influences policy choices, and aids in the defense of people, organizations, and society at large against cyberthreats.

Some of the cybersecurity study's objectives are as follows:

- To recognize the cybersecurity dangers and difficulties that individuals and companies face.
- To evaluate the performance of current cybersecurity measures and pinpoint opportunities for development.
- To research how cyber-attacks affect businesses, people, and the economy.
- To assess new technological developments and their possible influence on cybersecurity.
- To suggest methods and best practices for enhancing individual and corporate cybersecurity.

Prevention

Prevention is always better than cure. A multi-layered strategy that incorporates technological controls, user education, and proactive security measures is needed to prevent cyber-attacks. Here are some crucial actions people and organizations may take to reduce the danger of cyber-attacks. It is always advisable to be cautious when using the internet:

- To protect yourself from virus attacks, always use the most recent and updated antivirus software.
- Using firewalls may be advantageous.
- To prevent fraud, never send your credit card information to an unsecured website.
- **Maintain Software and Systems Updates:** To make sure you have the most recent security patches, update operating systems, software programs, and firmware on a regular basis. To speed up the process, enable automatic updates wherever feasible.
- Create complicated, strong passwords for each account, and refrain from using the same password across several platforms. To generate and store passwords safely, think about utilizing password managers.
- **Multi-factor Authentication (MFA) Should Be Used:** Wherever possible, enable MFA to offer an additional degree of security. To access an account, MFA requires users to submit several kinds of authentication, such as a password and a one-of-a-kind code texted to their mobile device.
- Use safe browsing techniques by exercising caution when opening attachments or links in emails, instant messaging, or strange websites. Make sure that websites are legitimate and that secure HTTPS connections are being used.
- **Use Reliable Security Software:** All devices should be equipped with up-to-date antivirus, anti-malware, and firewall software. These technologies aid in the detection and blocking of recognized dangers.
- Follow the 5Ps of online security: Precaution, Prevention, Protection, Preservation, and Perseverance [8].
- These are some of the preventions to keep us secure from cyber-attacks over the internet. Individuals and organizations may considerably lower their chance of being the target of cyber-attacks by adhering to these procedures and being watchful. Staying ahead of new threats in cybersecurity is a continuous process that requires ongoing assessment and adaptation of security measures.

Table 6.1 describes some of the incidents/cyber threats.

Cybersecurity risks are a serious worry for businesses and individuals all around the world. The threat environment is always changing, with new threats and attack methods appearing on a regular basis.

Table 6.1 Incidents/cyber threats till 2023

Cybersecurity threats	Date	Incident
SolarWinds Hack	December 2021	Hackers gained access to SolarWinds' IT management software, allowing them to conduct espionage on multiple US government agencies and other organizations.
Facebook Data Breach	April 2021	Personal data of over 500 million Facebook users was leaked online, including names, phone numbers, and other information.
Apple Zero-day Exploit	September 2021	Hackers exploited vulnerability in Apple's iMessage to install spyware on iPhones, affecting journalists, activists, and other targets.
Microsoft Exchange Server Hacks	January–March 2021	State-sponsored hackers exploited vulnerabilities in Microsoft Exchange Server to conduct espionage and steal data from businesses and organizations.
Inside Slack's Github Account Hack	December 29, 2022	Slack, one of the most popular business communication tools, became victim to a hacker.
Microsoft Azure Services Vulnerable to SSRF	January 17, 2023	Microsoft Azure services were vulnerable to server-side request forgery (SSRF) attacks due to four vulnerabilities.

Annual threat numbers are essential for better understanding the nature and effect of cybersecurity threats. The data gives significant insights into the most prevalent forms of attacks, the businesses and countries most impacted, and the expenses involved with cybercrime.

In this research chapter, we will review the yearly threat data since 2022, highlighting the most relevant trends and insights. We will also discuss the consequences of these figures for businesses and people, as well as tips for strengthening cybersecurity defenses (Figure 6.2).

Overview of cybersecurity

Cybersecurity is the activity of preventing unauthorized access, theft, or damage to electronic equipment, networks, and data. As we become more dependent on technology in our daily lives, cybersecurity has emerged as a crucial concern for all parties—people, businesses, and governments.

Viruses, malware, phishing schemes, ransomware, and DoS attacks are some examples of cyber threats existing nowadays. By interfering with business processes, stealing confidential information, and harming reputations, these dangers can seriously harm enterprises [11].

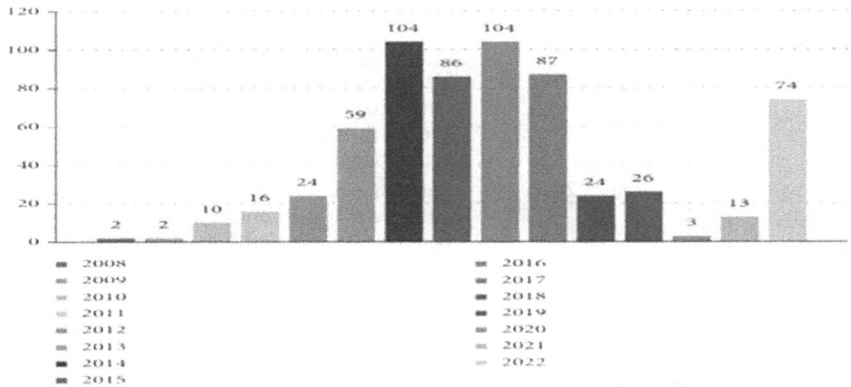

Figure 6.2 Threat intelligence statistics [10].

Organizations adopt a variety of cybersecurity solutions, such as firewalls, IDS, encryption, and routine security audits, to guard against these risks.

Cybersecurity is a rapidly evolving field, with cybercriminals constantly developing new methods to exploit weaknesses within networks and computers. It is critical for individuals and organizations to stay current on the latest cybersecurity trends and best practices in order to protect themselves from potential threats.

The following are the main objectives of cybersecurity:

Protect Confidentiality: Make sure that only authorized people or systems can access and see sensitive data and information, protecting it from unauthorized access.

Maintain Integrity: Prevent unauthorized alteration, tampering, or corruption to guarantee the precision, dependability, and consistency of data and systems.

Ensure Availability: Make sure that authorized users can access and utilize systems, networks, and data as necessary, preventing interruptions or downtime brought on by cyber-attacks.

Maintain Privacy: To avoid unauthorized disclosure or improper use of personal information, safeguard private and sensitive information by abiding by privacy laws and regulations.

Important elements of cybersecurity

- **Network Security:** Focusing on protecting networks and network infrastructure, network monitoring tools, virtual private networks (VPNs), intrusion detection and prevention systems (IDS/IPS), and firewalls are some examples of network security.

- **Application Security:** This entails protecting software applications at every stage of development, finding and fixing flaws, and using secure coding techniques.
- **Endpoint Security:** IoT devices, laptops, desktop computers, cellphones, and other endpoints are all covered under endpoint security. This includes countermeasures like patch management, device encryption, and antivirus software.
- **Data Security:** This involves safeguarding data while it is in use (being processed or accessed by authorized users), while it is in transit (being sent over networks), and while it is at rest (being stored).
- **Access Management (IAM):** By controlling and managing user access to systems, apps, and data, Identity and Access Management (IAM) makes sure that only authorized users have access to the necessary resources.
- To detect, respond to, and recover from security events or cyber-attacks, incident response establishes systems and procedures. This helps to reduce damage and resume regular operations.
- **Security Education and Training:** Informs people on cybersecurity best practices, makes them more conscious of dangers and hazards, and gives them instruction on how to spot and handle possible security issues.
- **Threat Intelligence** is the process of obtaining and examining data on current and potential threats, exchanging knowledge and indications of compromise (IOCs), and using that information to improve defenses and response capabilities.

To meet the changing threat landscape, cybersecurity is a continuous process that calls for constant monitoring, evaluation, and adaptation. To defend against cyber-attacks and uphold a safe digital environment, people, organizations, governments, and cybersecurity experts are cooperating in this endeavor.

Techniques of cybersecurity

Nowadays, numerous cybersecurity techniques are used to secure computers, networks, and crucial information from cyber threats. Among the most commonly used techniques are the following:

Encryption is the process of converting data into a code or cipher that only authorized parties can read. This method is frequently used to safeguard sensitive information.

Firewalls: A firewall is a security device that monitors and regulates network traffic that is both in and out. Firewalls can be hardware or software-based. It is able to block unauthorized access and stop malware from propagating.

Authentication of Data: The papers we acquire must always be validated before downloading, which means they must be reviewed to confirm that they have not been altered and have originated from a reliable source. Typically, antivirus programs installed on the devices authenticate these documents. To protect the gadgets from infections, a dependable antivirus program is required.

Two-Factor Authentication provides far more protection than the conventional username/password combination. Something you've got and something you are aware of is the two parts that constitute two-factor authentication. With this authentication method, a person must successfully complete two authentication steps in order to gain access to a website or account [12].

In single-factor authentication, the "something you know" element was the password. The additional element, or "something you have," is the most important part. There are various options for the components you have, including tokens, smart cards, pins/tans, and biometrics [8].

IDS and IPS: These systems keep track of network traffic and look for unusual behavior. It can also be used to guard against attacks by preventing traffic from known harmful sources from entering the network.

Vulnerability Scanning is the process of determining weaknesses or vulnerabilities in a system or network. This can be accomplished by automated tools or manual testing, and it can aid in the identification of potential points of attack.

Malware Detection and Removal Tools are used to detect and remove malicious software from a computer system. This includes viruses, Trojan horses, and other forms of malware.

Data Backup and Recovery: Data backup and recovery are critical for ensuring data integrity.

Cybersecurity uses a variety of methods and tools to safeguard systems, networks, and data against online dangers.

Following are some typical strategies and its implementation.

Access control

- **Implementation:** Use MFA, biometrics, or other robust authentication technologies to confirm user identities.
- Assign rights based on user roles and responsibilities by using role-based access control (RBAC).
- Use access control lists (ACLs) to limit user or group permission-based access to sensitive resources.
- To track and stop unauthorized access attempts, use IDS or intrusion prevention systems (IPS).

Encryption

- **Implementation:** Use encryption techniques to safeguard data while it is in use, transit, and at rest.
- To secure data transfer across networks, use the Secure Sockets Layer/ Transport Layer Security (SSL/TLS) protocols.
- Protect sensitive data on storage devices by using file-level encryption or full-disk encryption (FDE).
- Use end-to-end encryption (E2EE) to secure end-to-end communications.

Firewalls

- **Implementation:** Use firewalls to monitor and manage incoming and outgoing network traffic throughout implementation.
- Set up firewalls to enforce security rules, thwart unauthorized access attempts, and prioritize network traffic.
- To prevent the spread of assaults and control any breaches, split networks into smaller, more isolated portions.

Security information and event management (SIEM)

- **Implementation:** Use SIEM tools to deploy sources of security logs and event data for collection and analysis.
- Use SIEM systems to identify possible security problems, detect and correlate security events, and provide real-time warnings.
- Utilize SIEM's automatic response features to launch incident response procedures or start remediation procedures.

User education and awareness

- **Implementation:** Run user and employee education and awareness campaigns on cybersecurity.
- Inform users of the best ways to maintain their passwords securely, to browse safely, to be aware of social engineering techniques, and to spot phishing scams.
- Regularly inform the public on new dangers, cybersecurity regulations, and incident response techniques.

These are just a few illustrations of cybersecurity strategies. The organization's security needs, risk profile, and resource availability will determine the precise implementation and mix of approaches. To create a strong defense against cyber-attacks, it's crucial to have a tiered, all-encompassing strategy to cybersecurity.

Purpose of the report

A cybersecurity report's purpose is to provide an overview of the organization's cybersecurity posture and to identify potential vulnerabilities, threats, and risks that could jeopardize the security of its systems and data.

The report's goal is to provide stakeholders like senior management, IT staff, and other decision-makers with an understanding of the organization's current cybersecurity state and recommendations for improving its security posture.

A summary of the organization's current security posture, a description of the potential risks and threats facing the organization, an analysis of the organization's existing security controls and their effectiveness, and a set of recommendations for improving the organization's security posture is typically included in the report.

Finally, the goal of a cybersecurity report is to provide actionable information. Insights will assist the organization in reducing the risk of cybersecurity incidents and in protecting its systems and data from unauthorized access, use, disclosure, modification, or destructive overview of the study of cybersecurity threats.

Cybersecurity is the practice of preventing unauthorized access, theft, damage, or destruction of computer systems, networks, and digital information. It is an important field in today's interconnected world, where cyber threats are becoming more sophisticated and common.

As technology evolves at an incredible rate, so are the threats we confront in cyberspace. Every year, new threats arise, and fraudsters discover new methods to exploit flaws in our systems and networks.

Some of these challenges and threats include:

Insider Threats: Insiders who have authority over sensitive data can be a major cybersecurity concern. Organizations must have policies and procedures in place to protect themselves from insider threats.

APTs are highly focused and sophisticated attacks that are hard to detect and prevent because they may evade typical security measures [11].

Man-in-the-Middle (MitM) Attacks: MitM attacks happen whenever an attacker monitors conversations between two parties in order to eavesdrop, steal data, or mimic one of them [13].

Supply Chain Attacks: Supply chain attacks involve gaining unauthorized access to systems and data by exploiting vulnerabilities in the supply chain.

Phishing: It is the practice of using emails to transmit malicious messages or social engineering. The purpose is to steal private data like debit card numbers and usernames and passwords from the victim. This assault is typically employed as part of a wider operation to establish a stronghold among government or company networks as an advance and persistent threat.

SQL Injection (SQLI): It tries to modify the back-end access to database information that was never meant to be displayed. SQL injections might be performed by simply typing harmful code into a susceptible website search field [14].

Lack of Skilled Professionals: Because of the shortage of competent cybersecurity specialists, it is challenging for organizations to identify and employ the expertise needed to defend their networks and systems.

DDoS Attacks: DDoS attacks refer to a particular form of attack in which several compromised systems work together to target a single victim, ultimately resulting in the denial of service for users attempting to access the targeted system (Figure 6.3).

To avoid these threats, cybersecurity experts must always be watchful and adapt their techniques to the changing situation. This necessitates requiring awareness of developing trends and prospective cybersecurity risks.

In this research chapter, we will investigate the developing cybersecurity trends that are expected to shape the sector in the coming years. Organizations may design more effective cybersecurity plans to secure their assets and reduce the risk of a cyber-attack by spotting these trends and understanding their consequences.

Some of the upcoming cybersecurity trends

Artificial Intelligence and Machine Learning: AI and machine learning can help cybersecurity professionals identify patterns in cyber-attacks, allowing them to predict and prevent attacks.

Cloud Security: Because cloud computing has become an essential component of modern business operations, it has become an appealing target for cybercriminals. As a result, cloud security will become more important [15].

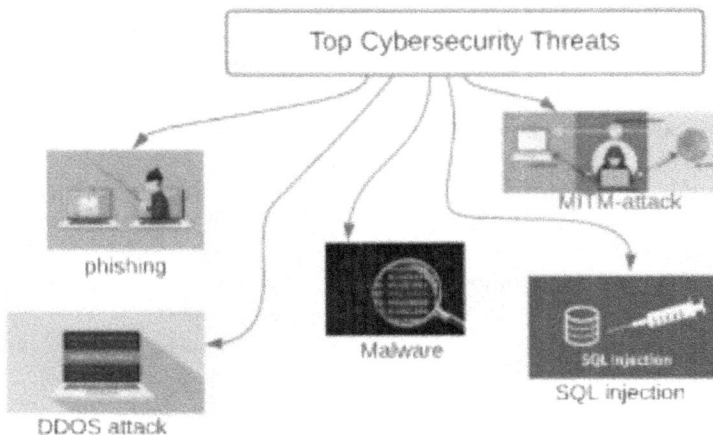

Figure 6.3 Top cybersecurity threats.

Quantum Computing: Use of this could result in the ability to compromise numerous encryption techniques that are presently employed to safeguard data, which could have serious implications for cybersecurity. As quantum computing becomes more widely available, new approaches to encryption will be required to ensure data security.

Zero Trust Security: This cybersecurity strategy in question assumes that any network traffic could be harmful, and therefore requires that all users and devices be authenticated and verified before they are allowed to access resources. This approach can be helpful in preventing cyber-attacks, including data breaches [15].

Privacy and Data Protection: Stricter legislation like the General Data Protection Regulation (GDPR) and the California Consumer Privacy Act (CCPA) will only increase the necessity of privacy and data protection. To secure user data, organizations will need to give privacy-by-design principles top priority, employ privacy-enhancing technology, and establish stricter data protection policies.

Ransomware Defense and Resilience: Disruptive and common ransomware assaults are on the rise. To lessen the effects of ransomware attacks, organizations will need to invest in comprehensive ransomware defense techniques, such as proactive threat hunting, network segmentation, secure backup systems, and incident response plans.

It's crucial to remember that these problems and risks might materialize depending on trends that have been seen in the years leading up to 2023. The cybersecurity environment is ever-changing, and as attackers modify their strategies, new risks may materialize. To effectively manage these difficulties, organizations and people should exercise vigilance, adopt best practices, and stay current with cybersecurity advancements.

However, along with these trends, there are also various challenges and threats that cybersecurity professionals need to address (Figure 6.4).

As technology evolves continuously, cybersecurity experts must stay current on the newest trends and dangers. Continuous training and education, as well as engagement with other cybersecurity specialists and organizations, are required. Furthermore, regulatory frameworks are growing more complicated, and organizations are facing compliance challenges due to regulations related to data privacy and security, such as the General Data Protection Regulation (GDPR) and the California Consumer Privacy Act (CCPA).

One of the key challenges is the lack of cybersecurity experts, which is a significant issue that is projected to persist in the upcoming years. Organizations must search for cutting-edge solutions to this problem, such as automation and outsourcing to bolster their cybersecurity teams.

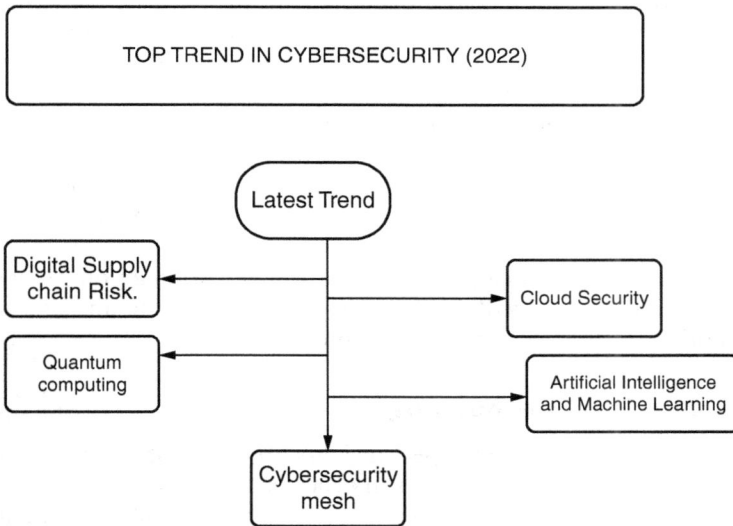

```
            ┌─────────────────────────────────────────┐
            │                                          │
            │   TOP TREND IN CYBERSECURITY (2022)       │
            │                                          │
            └─────────────────────────────────────────┘
```

Figure 6.4 Trends in cybersecurity 2022.

METHODOLOGY

The following steps are typically included in the methodology for addressing cybersecurity challenges and threats.

The research highlights investigations made in the area of cybersecurity threats, trends, and challenges. We searched several databases using terms like cybersecurity and challenges related to it [16].

Risk Assessment: Perform a comprehensive evaluation of the risks to detect possible weaknesses and dangers in the infrastructure, data, and systems of the organization. This can involve recognizing potential hazards, examining their probability, and determining their effect on the organization.

Identification: The first stage is to identify your organization's cybersecurity concerns, trends, and threats. This can be accomplished through the evaluation of threat intelligence reports, the analysis of attack trends, and the conduct of risk assessments.

Create a Security Strategy: Based on the risk assessment results, create a comprehensive security strategy that includes a mix of hardware and software solutions, policies and procedures, and employee training and education.

Implementation: This entails putting in place the security strategy, which includes hardware and software solutions, policies and procedures, and employee training and education. This includes the installation of firewalls, antivirus software, IDS, and other security measures.

Ongoing Monitoring and Evaluation: Conduct ongoing monitoring and evaluation. Monitoring and evaluation are performed to ensure that the security strategy is effective and up-to-date. Regular vulnerability assessments, penetration testing, and incident response planning can all be part of this.

Continuous Improvement: Improve the security strategy on an ongoing basis by incorporating feedback from ongoing monitoring and assessment and adapting to changes in the threat landscape.

Overview of the methodology

Research, data analysis, and expert insights are frequently used to analyze emerging trends, issues, and threats in cybersecurity. The methodology frequently used to examine the shifting cybersecurity landscape is summarized as follows.

Research and Literature Review: To obtain insight into historical and present cybersecurity trends, issues, and threats, researchers examine existing literature, academic papers, industry reports, and case studies. By doing so, it is possible to build a baseline understanding and spot knowledge gaps.

Data Gathering: Data is gathered from a variety of sources to learn more about incidents, vulnerabilities, and cyber-attacks. Analyzing information from security vendors, threat intelligence feeds, incident reports, and governmental sources is part of this process. Additionally, surveys, questionnaires, or interviews with specialists in the field of cybersecurity and business leaders can be used to gather the data.

Data Analysis: To find patterns, trends, and correlations, the obtained data is analyzed using a variety of statistical and analytical approaches. This study aids in the discovery of new risks, weaknesses, and attack vectors as well as their potential effects on people, organizations, and digital infrastructure.

Collaboration with cybersecurity specialists, businesspeople, and researchers is essential to gaining knowledge about the most recent advancements and upcoming difficulties. Experts give subject-specific expertise, viewpoints on present and potential dangers, and assistance with peer review to confirm results.

Threat Modeling: Assessing possible threats and vulnerabilities in a system or network is the process of "threat modeling." Researchers can create mitigation plans and countermeasures by identifying possible attack routes and comprehending the intentions and capabilities of potential attackers.

Scenario Analysis and Simulations are used by researchers to gauge the effects of future cyber-attacks and gauge how well current security measures are working. This aids in locating weak points, enhancing incident response procedures, and creating preventative defense tactics.

LITERATURE REVIEW

In our review of the literature, we examined numerous academic fields, such as computer science, engineering, political studies, psychology, security studies, management, education, and sociology. Our findings indicated that the most prevalent fields in our review were engineering, technology, computer science, security, and defense. Nonetheless, we also found limited instances of cybersecurity being discussed in journals related to policy development, law, healthcare, public administration, accounting, management, sociology, psychology, and education.

Studies in the literature on current threats, trends, and difficulties in cybersecurity offer insightful information on how the field is changing. Researchers and practitioners may keep up with the latest developments and pinpoint prospective areas of emphasis by reading academic papers, industry reports, and expert analysis.

A few of the literature reviews we encountered regarding cybersecurity are as follows:

Skilled Cybersecurity Professionals: The dearth of proficient cybersecurity experts is a major issue that is likely to persist in the upcoming years. As per a report released by the ISC, the scarcity of qualified cybersecurity professionals around the world is projected to surge to 2.1 million by the year 2023 [17].

The Complexity of Modern Cybersecurity Systems: The complexity of modern cybersecurity systems is a significant challenge for organizations, as they require a range of hardware and software solutions to protect against a variety of threats. This complexity can make it difficult for organizations to implement and maintain effective cybersecurity strategies.

Rapidly Evolving Nature of Cyber Threats: Organizations find it challenging to keep up with the ever-evolving cyber threats. The average cost of a data breach, as per a Ponemon Institute report, is $3.86 million, which escalates for organizations that encounter breaches due to new or emerging threats.

Ransomware Attack: Cyber-attacks involving ransomware are on the rise, and entities of all sizes are vulnerable to such attacks. According to Cybersecurity Ventures, the global cost of ransomware damage is expected to reach $32 billion by 2023.

Here are some key areas typically covered in literature studies:

Threat Environment Analysis:

The threat environment as it exists now is examined in the literature, along with the many cyberthreats, attack methods, and approaches used by cybercriminals.

They look at the motives behind cyber-attacks, including monetary gain, political goals, espionage, or service interruption.

Researchers investigate the most recent developments in cybercrime, such as ransomware, phishing, social engineering, insider threats, advanced persistent threats (APTs), and assaults on vital infrastructure, IoT devices, and cloud infrastructure.

Deficiencies and Threats:

Studies in literature examine security holes in network protocols, hardware, and software.

They examine the possible effects on systems and data as well as how attackers exploit these vulnerabilities.

Researchers examine the procedures for disclosing vulnerabilities, ethical disclosure standards, and the efficiency of patch management in resolving vulnerabilities.

Threats from Emerging Technologies:

Studies in the literature look at how upcoming technologies like cloud computing, blockchain, 5G, IoT, and machine learning will affect security [18–21].

They evaluate the possible dangers, weaknesses, and difficulties linked to implementing these technologies.

To adopt and secure new technologies, researchers investigate security aspects and best practices.

Researchers get a complete grasp of cybersecurity concerns, emerging trends, and developing threats by conducting literature studies on these and other relevant subjects. This knowledge is crucial for building successful defense tactics, influencing policy choices, and leading the development of novel cyber risk-mitigation technologies and practices.

CONCLUSION

As the world becomes more interconnected, networks are being used to carry out critical transactions, making computer security an increasingly important and broad topic.

Cybercrime is constantly evolving, with new technologies and threats emerging every day, straining organizations' infrastructure security efforts. To reduce cybercrime, a proactive approach to cybersecurity is recommended, including implementing strong security measures such as firewalls, antivirus software, and IDS. It is also important to educate staff members about cyber threats and train them to spot and report any suspicious activity.

The use of cloud computing and the IoT has introduced new challenges for cybersecurity due to their complexity, lack of standardization, and lack of regulation. While cybercrimes cannot be eliminated, efforts should be made to reduce them to ensure a safe and secure future in cyberspace [22].

In conclusion, cybersecurity challenges, ongoing threats, and new trends continue to evolve, requiring organizations to stay vigilant and adapt their security strategies accordingly.

Studying cybersecurity threats, trends, and challenges highlights how crucial it is to approach cybersecurity in a proactive manner. Organizations must constantly evaluate their security posture, pinpoint vulnerabilities, and implement precautionary measures. They are able to better defend their systems, data, and reputation from the constantly changing threat environment by doing this.

The overall study concludes that cybersecurity is a dynamic, complex field that needs ongoing care and funding. Organizations can reduce their risk and protect their assets from cyber threats by following best practices, staying informed about emerging threats and trends, and investing in effective security solutions.

REFERENCES

1. Das, S. (2021). Adequacy and limitations of the information technology act in addressing cyber-security issues of Indian power systems. In *IEEE International Conference on Power Systems (ICPS)*.
2. Stevenson, C. L. (1908–1979). Analytic philosophe (defining cybersecurity).
3. Craigen, D. (2014). Defining cybersecurity. *Technology Innovation Management Review*.
4. Anand, R., Shrivastava, G., Gupta, S., Peng, S. L., & Sindhwani, N. (2018). Audio watermarking with reduced number of random samples. In *Handbook of Research on Network Forensics and Analysis Techniques* (pp. 372–394). IGI Global.
5. Sindhwani, N., Maurya, V. P., Patel, A., Yadav, R. K., Krishna, S., & Anand, R. (2022). Implementation of intelligent plantation system using virtual IoT. *Internet of Things and Its Applications*, 305–322.
6. Anand, R., Sindhwani, N., & Juneja, S. (2022). Cognitive Internet of Things, its applications, and its challenges: A survey. In *Harnessing the Internet of Things (IoT) for a Hyper-Connected Smart World* (pp. 91–113). Apple Academic Press.
7. Kaur, J., Sindhwani, N., Anand, R., & Pandey, D. (2022). Implementation of IoT in various domains. In *IoT Based Smart Applications* (pp. 165–178). Cham: Springer International Publishing.
8. Dashora, K. (2011). Cyber crime in the society: Problems and preventions. *Journal of Alternative Perspectives in the Social Sciences*.
9. Kantaria, G. (2020). Cyber threats and attack vectors during COVID-19. *International Journal for Research in Applied Science and Engineering Technology*, no. ijraset.2020.32013.

10. Chen, S. (2022). An automatic assessment method of cyber threat intelligence combined with ATT&CK matrix. *Wireless Communications and Mobile Computing.*

11. Hussain, A., Mohamed, A., & Razali, S. (2020, March). A review on cybersecurity: Challenges & emerging threats. In *Proceedings of the 3rd International Conference on Networking, Information Systems & Security* (pp. 1–7).

12. Mogal, M. M. (2022). How two factor authentication helps in cybersecurity. *International Research Journal of Modernization in Engineering Technology and Science*, 4(06).

13. Sajal, S. Z., Jahan, I., & Nygard, K. E. (2019). A survey on cyber security threats and challenges in modern society. In *2019 IEEE International Conference on Electro Information Technology (EIT)*. IEEE.

14. Tsochev, G., Trifonov, R., Nakov, O., Manolov, S., & Pavlova, G. (2020). Cyber security: Threats and challenges. In *2020 International Conference Automatics and Informatics (ICAI)* (pp. 1–6), Varna, Bulgaria. doi: 10.1109/ICAI50593.2020.9311369.

15. A Sophos Article 04. 12v1.dNA, eight trends changing network security by James Lynne.

16. Rahman, N. A. A., Sairi, I. H., Zizi, N. A. M., & Khalid, F. (2020). The importance of cybersecurity education in school. *International Journal of Information and Education Technology.*

17. Nafea, R. A., & Amin Almaiah, M. (2021). Cyber security threats in cloud: Literature review. In *2021 International Conference on Information Technology (ICIT)* (pp. 779–786), Amman, Jordan. doi: 10.1109/ICIT52682.2021.9491638.

18. Sindhwani, N., Anand, R., Vashisth, R., Chauhan, S., Talukdar, V., & Dhabliya, D. (2022, November). Thingspeak-based environmental monitoring system using IoT. In *2022 Seventh International Conference on Parallel, Distributed and Grid Computing (PDGC)* (pp. 675–680). IEEE.

19. Anand, R., Ahamad, S., Veeraiah, V., Janardan, S. K., Dhabliya, D., Sindhwani, N., & Gupta, A. (2023). Optimizing 6G wireless network security for effective communication. In *Innovative Smart Materials Used in Wireless Communication Technology* (pp. 1–20). IGI Global.

20. Jain, S., Sindhwani, N., Anand, R., & Kannan, R. (2022, March). COVID detection using chest X-ray and transfer learning. In *Intelligent Systems Design and Applications: 21st International Conference on Intelligent Systems Design and Applications (ISDA 2021) Held During December 13–15, 2021* (pp. 933–943). Cham: Springer International Publishing.

21. Shukla, R., Dubey, G., Malik, P., Sindhwani, N., Anand, R., Dahiya, A., & Yadav, V. (2021). Detecting crop health using machine learning techniques in smart agriculture system. *Journal of Scientific & Industrial Research*, 80(08), 699–706.

22. Pandey, A. B., Tripathi, A., & Vashist, P. C. (2022). A survey of cyber security trends, emerging technologies and threats. *Cyber Security in Intelligent Computing and Communications*, 19–33.

Chapter 7

Intelligence approaches for products, operations, systems, and services

Anil Kumar Dubey, Vikash Yadav, and Chuan-Ming Liu

INTRODUCTION

In today's fast-changing technological environment, intelligence approaches—often referring to the incorporation of artificial intelligence (AI) and data-driven methodologies into different elements of products, operations, systems, and services—have grown in significance (Figure 7.1).

Here are some examples of how these techniques can be used in each of these fields.

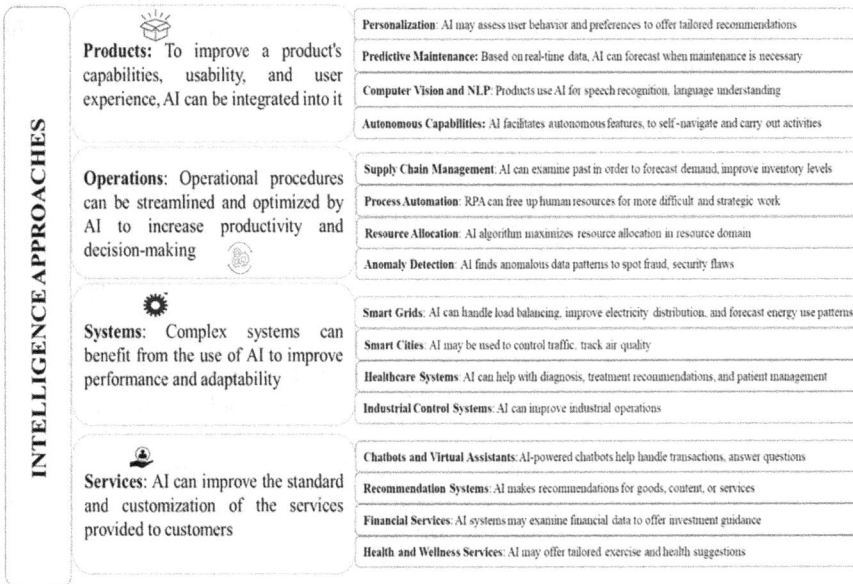

Figure 7.1 Intelligent approaches for product, operation, system, and services.

DOI: 10.1201/9781003479031-7

Products: To improve a product's capabilities, usability, and user experience, AI can be integrated into it. Several methods include the following:

- **Personalization:** In goods like streaming services, e-commerce platforms, and social media, AI may assess user behavior and preferences to offer tailored recommendations and suggestions.
- **Predictive Maintenance:** Based on real-time data, AI can forecast when maintenance is necessary for industrial products like machinery and equipment, minimizing downtime and increasing productivity.
- **Computer Vision and Natural Language Processing:** Products can use AI to do tasks like speech recognition, language understanding, and image recognition. This can improve user interfaces and interactions with products.
- **Autonomous Capabilities:** Autonomous features, like those in self-driving cars, drones, or robotic vacuum cleaners, can be found in AI-driven goods, allowing them to navigate and carry out activities without direct human direction.

Operations: Operational procedures can be streamlined and optimized by AI to increase productivity and decision-making. Several methods include the following:

- **Supply Chain Management:** To forecast demand, improve inventory levels, and suggest supply chain solutions, AI can examine past data.
- **Process Automation:** By automating routine operations, robotic process automation (RPA) can free up human resources for more difficult and strategic work.
- **Resource Allocation:** In domains like energy distribution, workforce scheduling, and financial portfolio management, AI algorithms can maximize resource allocation.
- **Anomaly Detection:** AI can find anomalous data patterns to spot fraud, security flaws, or flaws in intricate systems.

Systems: Complex systems can benefit from the use of AI to improve performance and adaptability. Several methods include the following:

- **Smart Grids:** In smart grid systems, AI can handle load balancing, improve electricity distribution, and forecast energy use patterns.
- **Smart Cities:** AI may be used to control traffic, track air quality, choose the best routes for collecting waste, and generally improve urban management.
- **Healthcare Systems:** In healthcare systems, AI can help with diagnosis, treatment recommendations, and patient management, improving patient outcomes.

- **Industrial Control Systems:** AI can improve industrial operations, keep track of equipment health, and improve safety.

Services: AI can improve the standard and customization of the services provided to customers. Several methods include the following:
- **Chatbots and Virtual Assistants:** AI-powered chatbots help handle transactions, answer questions, and offer real-time customer service.
- **Recommendation Systems:** Engines powered by AI make recommendations for goods, content, or services based on user interests and behavior.
- **Financial Services:** AI systems may examine financial data to offer investment guidance, determine credit risk, and spot fraud.
- **Health and Wellness Services:** Based on user data from wearable's and applications, AI may offer tailored exercise and health suggestions.

The incorporation of intelligence approaches into operations, systems, and services calls for thorough domain knowledge, data accessibility, moral considerations, and constant optimization. The objective is to apply AI to develop more effective, flexible, and user-centric solutions.

PRODUCTS

AI and data-driven techniques are used in intelligence approaches for products to improve their features, capabilities, and overall worth. These strategies may result in better performance, more personalization, and enhanced user experiences. To process large amounts of real-time data for the industry environment, notably for a product or a set of products in the online e-commerce industry, soft intelligence [11] approaches reduce the amount of time needed to find the best solution from a large solution base. AI must be used to harness its capacity to restructure businesses to meet changing consumer demands if they are to survive in the highly competitive global business climate. Here are some crucial product intelligence techniques:

Personalization: To offer individualized recommendations and experiences, AI may analyze user data and behavior. Examples include individualized content suggestions on streaming platforms, specific product recommendations on e-commerce websites, and individualized user interfaces in software programs.

Predictive Analytics: AI has the ability to forecast future patterns and behaviors using historical data. This allows for proactive steps like stocking up on popular things by anticipating client preferences, product demand, and even future product problems [6].

Natural Language Processing (NLP) and Voice Recognition: The ability to interpret and respond to natural language commands and inquiries is made possible by integrating NLP with voice recognition technology. Voice assistants like Apple's Siri and Amazon's Alexa, which can operate smart home gadgets, respond to queries, and carry out tasks based on voice commands, are examples of this.

Computer Vision: Products can grasp and interpret visual data thanks to computer vision. This is utilized in augmented reality applications that superimpose digital data on the physical world, object detection for autonomous vehicles, and facial recognition for device unlocking.

Autonomous Capabilities: AI-powered autonomous features enable items to carry out activities without direct human control. Autonomous vehicles such as self-driving automobiles, drones, and robotic vacuum cleaners are some examples.

Quality Control and Anomaly Detection: AI may examine sensor data to find flaws or irregularities in items as they are being manufactured. This enhances quality control and guarantees that only goods fulfilling particular standards are sold.

Sentiment Analysis: To determine customer sentiment and satisfaction, AI can evaluate user reviews and feedback. It is possible to use this data to enhance and improve products.

Virtual Reality (VR) and Augmented Reality (AR): By constructing immersive worlds or presenting digital information on top of the actual world, these technologies can improve user experiences. This can be utilized in a variety of goods, including those for real estate and gaming.

Product Lifecycle Management: A product's full lifecycle, from design and development to production and service, may be managed with the help of AI. It can streamline procedures, spot bottlenecks, and boost effectiveness at every level.

Smart Sensors and Internet of Things (IoT): Data collection and real-time monitoring are made possible by integrating smart sensors into devices and connecting them to the IoT. With the help of this information, products may be made to perform better, maintenance issues can be found, and product improvements can be made.

Data protection, ethical ramifications, and user consent must all be carefully considered when developing intelligence-based product strategies. Companies may develop products that are more user-friendly, effective, and responsive by utilizing AI technologies.

EXAMPLE OF INTELLIGENCE APPROACHES FOR PRODUCTS

Here are a few instances of products that have been given intelligence approaches:

Smart Home Devices: Thermostats with AI capabilities, such as the Nest Learning Thermostat, are an example of products in the smart home market. It optimizes energy usage and comfort by learning user preferences and adjusting the temperature accordingly. Another illustration is the Amazon Echo, which listens to voice requests, manages smart devices, and offers tailored recommendations depending on user interactions.

Fitness Wearables: AI is used by fitness trackers like the Apple Watch and Fitbit to monitor users' heart rates, physical activity, sleep habits, and other data. These tools offer individualized health information, such as workout recommendations depending on the user's objectives and current fitness level.

E-commerce Recommendations: AI-driven recommendation systems are used by platforms like Netflix and Amazon to offer products or content to users. While Netflix recommends movies and TV episodes based on viewing history and preferences, Amazon bases its product suggestions on a user's browsing history and purchasing behavior.

Autonomous Vehicles: A variety of AI techniques, such as computer vision and machine learning, are used in the development of self-driving cars, such as those being produced by businesses like Tesla and Waymo. Without human assistance, these cars can sense their surroundings, make quick decisions, and travel securely.

Smart Cameras: Security cameras with AI capabilities, like those made by businesses like Ring, can discriminate between people, animals, and other objects, minimizing false alarms. Additionally, they can deliver real-time video streams and transmit notifications to customers' smartphones.

Language Translation Devices: Real-time language translation is possible with AI-driven translation tools like Google's Pixel Buds, enabling users to converse with persons who speak different languages.

Augmented Reality Apps: Using the camera of a smartphone, apps like Pokémon GO use AR to superimpose digital objects—in this example, Pokémon characters—on top of the real environment. Users are given an interactive and immersive experience as a result of this.

Smart Appliances: AI-enabled gadgets, such as smart refrigerators, may keep an eye on food supplies, track expiration dates, and make recipe recommendations based on the goods on hand. These devices can gradually pick up on human preferences.

Health Monitoring Devices: Wearable medical equipment, such as glucose monitors for diabetics, can continuously check blood sugar levels and send real-time notifications when levels go outside of the appropriate range.

Gaming Consoles: AI techniques are used by current gaming consoles like the PlayStation 5 and Xbox Series X to improve gameplay, visuals, and load times.

These instances show how applying intelligence to products may improve them by delivering individualized experiences, automation, real-time insights, and enhanced functionality.

OPERATIONS

AI and data-driven techniques are used in intelligence approaches for operations in order to expedite and improve a variety of operational processes within enterprises. These strategies seek to boost decision-making, increase efficiency, and lower costs. Operational intelligence (OI) is a kind of data analysis that allows decisions and actions to be based on real-time data as it is generated or gathered by businesses. The data analysis process is typically automated, and the information obtained is then linked into operational systems for quick access by business management and employees. OI apps are generally intended for front-line employees who, if given timely business intelligence (BI) and analytics data, should be able to make better business decisions or act more quickly on problems. Agents at call centers, salespeople, internet marketing groups, logistics planners, manufacturing managers, and medical personnel are a few examples. OI can also be utilized to automatically launch responses to specific events or circumstances. Here are a few instances of intelligence-gathering techniques for operations:

Supply Chain Management: To forecast demand trends, optimize inventory levels, and advise effective supply chain tactics, AI can examine historical data and outside influences. This may result in fewer stockouts, less overstocking, and overall greater supply chain effectiveness.

Process Automation: Software robots are used in RPA to automate repetitive operations including data entry, invoice processing, and report preparation. Employees can now concentrate on more important and valuable work because of the decrease in human error.

Predictive Maintenance: AI can evaluate sensor data to forecast when machinery or equipment is likely to malfunction in sectors like manufacturing [3, 4, 5, 9, 10] and transportation. Because of this, businesses can execute maintenance prior to a failure, limiting downtime and lowering maintenance expenses.

Resource Allocation: In domains like labor scheduling, energy distribution, and financial portfolio management, AI algorithms can maximize resource allocation. This guarantees effective and efficient utilization of resources.

Anomaly Detection: To spot anomalies or strange trends, AI can continuously analyze massive amounts of data. AI can recognize possible threats or breaches in fields like cybersecurity [7, 8] and take proactive steps to reduce risks.

Quality Control and Inspection: Using computer vision and picture recognition, AI-powered systems can evaluate products and find flaws. To guarantee product quality, this is especially helpful in the industrial sector.

Energy Management: AI can analyze data from sensors and smart meters to optimize energy use in buildings and other facilities. This may result in energy savings and lower operating expenses.

Healthcare Operations: Acuity-based resource allocation and patient flow optimization are all possible in hospitals with AI. This improves the quality of patient treatment and raises the general effectiveness of healthcare operations.

Financial Operations: Financial data may be analyzed by AI algorithms to spot fraudulent activity in real time, increasing the security of financial transactions. AI can also automate standard financial procedures like reconciliation and invoicing.

Logistics and Transportation: Route planning for delivery vans can be made more efficient by AI by taking into account variables like traffic and weather forecasts. Deliveries become quicker and more effective as a result.

Project Management: Tools for project management powered by AI can help with work scheduling, resource allocation, and risk assessment, which will improve project deadlines and results.

Customer Service and Support: Virtual assistants and chatbots powered by AI can help users troubleshoot issues, respond to frequently asked inquiries, and offer real-time customer care.

By enabling data-driven decision-making, automating procedures, and optimizing resource allocation, intelligence approaches for operations can transform the way organizations run, ultimately enhancing productivity and competitiveness.

EXAMPLE OF INTELLIGENCE APPROACHES FOR OPERATION

Some examples of intelligence approaches applied to operations:

Predictive Maintenance in Manufacturing: AI can use sensor data from equipment to analyze manufacturing plants and forecast when maintenance is required. By planning maintenance before a failure occurs, this reduces unplanned downtime and increases production efficiency.

Chatbot Customer Support: To offer real-time customer help, many businesses use chatbots powered by AI on their websites. These chatbots may help clients through the buying process, address frequent queries, and fix problems, all of which improve customer happiness.

Dynamic Pricing in E-commerce: On the basis of variables including demand, competition, and user behavior, e-commerce platforms frequently use AI algorithms to dynamically change product prices. Providing the correct pricing at the right moment enables businesses to maximize their revenue.

Energy Management in Buildings: AI is used by smart building systems to assess energy usage trends and modify the lighting, heating, and cooling systems as necessary. This leads to energy savings while preserving the highest levels of comfort for the residents.

Inventory Optimization: Retailers can use AI to examine past sales data and forecast product demand. This assists in maximizing inventory levels, lowering carrying costs, and preventing stockouts.

Fraud Detection in Banking: Financial organizations use AI algorithms to quickly identify fraudulent transactions. To identify suspicious activity and evaluate transaction patterns, these algorithms aid in preventing financial losses.

Optimized Route Planning: Utilizing real-time traffic information, weather forecasts, and delivery window information, logistics and delivery companies employ AI to improve delivery routes. This lowers fuel expenses and increases delivery effectiveness.

Personnel Scheduling in Healthcare: AI can be used by hospitals to schedule medical personnel based on patient demand, staff availability, and necessary skill sets. This ensures effective patient care and sufficient personnel levels.

Quality Control in Manufacturing: On the production line, goods are examined for flaws and departures from quality requirements using AI-driven computer vision systems. This lessens the necessity for human examination and raises the standard of the produce.

Risk Assessment in Insurance: AI is used by insurance companies to evaluate risks and calculate insurance rates. To estimate the possibility of claims, AI systems examine multiple data points, assisting insurers in making strategic decisions.

Asset Tracking and Management: Asset location and condition can be tracked in real time by tracking systems using AI. This is helpful in sectors like logistics, where it's crucial to trace the transit of items.

Employee Performance Analysis: Data about employee performance can be analyzed by AI to find patterns, areas of strength, and areas for development. With the use of this knowledge, companies can offer specialized chances for training and growth.

These examples demonstrate how intelligence techniques can be applied to diverse operational contexts to enhance effectiveness and efficiency.

SYSTEM

Intelligent systems are highly developed machines that can perceive their environment and react to it. Intelligent devices come in a variety of shapes and sizes, from robotic vacuums like the Roomba to face recognition software to Amazon's customized shopping recommendations. To optimize and improve the performance of complex systems, intelligence approaches for systems integrate AI and data-driven strategies. These strategies could include smart grids, healthcare systems, and smart cities. Here are some instances of system intelligence strategies:

Smart Grids: AI is used in smart grids to continuously monitor and control how electricity is distributed. For effective energy flow optimization, power outage reduction, and integration of renewable energy sources, AI algorithms analyze data from sensors and meters.

Healthcare Systems: To help with diagnosis, treatment planning, and patient care, AI can be included in electronic health records. AI, for instance, may examine X-rays and MRI pictures to help doctors identify disorders.

Smart Cities: Smart cities may monitor traffic patterns, air quality, trash management, and energy use via AI-driven technologies. Utilizing this data will improve urban planning, lessen traffic, and improve citizens' quality of life in general.

Industrial Control Systems: AI may be used in industrial settings to improve manufacturing processes, keep track of equipment health, and guarantee worker safety. AI, for instance, can evaluate sensor data to forecast equipment breakdowns and initiate maintenance procedures.

Financial Trading Systems: To spot trading patterns, predict market trends, and place transactions at the right moments, AI systems can scan enormous volumes of financial data. This is utilized in algorithmic trading systems.

Agricultural Systems: To keep track of crop health, weather patterns, and soil conditions, AI can analyze data from sensors and satellites. Farmers can use this knowledge to make educated decisions about irrigation, fertilizer, and pest control.

Air Traffic Control: By examining real-time data on flight paths, weather, and airport operations, AI can help air traffic controllers. By doing so, air traffic flow is optimized, and congestion is avoided.

Water Management Systems: In water management systems, AI can forecast water demand, monitor water quality, and optimize water distribution. This is especially important to guarantee effective water usage in urban and rural settings.

Environmental Monitoring: To track environmental parameters like deforestation, pollution levels, and wildlife populations, AI can interpret data from distant sensors and satellites. This knowledge helps conservation efforts.

Disaster Management: By evaluating data from a variety of sources to predict disaster trends, evaluate risks, and effectively allocate resources, AI can assist in managing and responding to natural disasters.

Educational Systems: AI can be used in educational systems to tailor learning experiences for students, spot areas where they may require more support, and modify lesson plans in accordance with student success [1, 2].

Telecommunication Networks: By examining data on user behavior, network traffic, and infrastructure health, AI can improve network performance. By doing this, reliable communication services are ensured.

These examples show how intelligence-based approaches can be used to improve various systems' efficacy, safety, and overall efficiency. Resource allocation, decision-making, and adaptability can all be improved when AI is incorporated into complex systems.

EXAMPLES OF INTELLIGENCE APPROACHES FOR SYSTEM

Some examples of intelligence approaches applied to different types of systems:

Smart Grids: To improve energy distribution in a smart grid system, AI algorithms can examine real-time data from energy-producing sources, consumption trends, and weather forecasts. For instance, the system may manage load shedding and effectively allocate energy resources at times of peak demand to avoid blackouts.

Healthcare Systems: AI can help with diagnosis and therapy suggestions in the medical field. For instance, IBM's Watson for Oncology employs AI to examine clinical trial data, patient information, and medical literature to assist physicians in selecting the best course of treatment for individual cancer patients.

Smart Cities: Utilizing AI to control traffic is one intelligence strategy in a smart city. An AI-based traffic management system, for instance, can evaluate real-time traffic data, modify the timing of traffic signals, and recommend other routes to ease congestion and enhance traffic flow.

Industrial Control Systems: AI can optimize production processes in industrial automation. AI, for instance, may evaluate sensor data to forecast machine faults and optimize production schedules to reduce downtime in the manufacturing industry.

Financial Trading Systems: Algorithms are used by AI-driven trading platforms to evaluate market movements and carry out trades quickly. With the help of these tools, traders can quickly and intelligently spot trends in financial data.

Agricultural Systems: Agriculture irrigation systems can be improved with AI. AI algorithms can decide the best amount and timing of irrigation to maximize crop output and preserve water by examining soil moisture levels, weather forecasts, and crop characteristics.

Air Traffic Control: AI-enhanced air traffic control systems can redirect flights to reduce delays and increase airspace usage. They can also forecast air traffic congestion. Additionally, these tools help controllers manage air traffic flow more effectively.

Water Management Systems: Based on past data, weather predictions, and usage trends, AI can forecast water demand. In water management systems, this information can be utilized to regulate reservoir levels and optimize water distribution.

Environmental Monitoring: AI can track rates of deforestation and identify illicit logging activity in conservation initiatives by analyzing satellite imagery. This knowledge enables government agencies to respond quickly to safeguard forests and wildlife.

Disaster Management: To forecast and evaluate the effects of natural disasters, AI-powered systems may analyze data from a variety of sources, including seismic sensors and meteorological forecasts. This makes it possible for governments and aid organizations to deploy resources wisely in times of need.

These instances show how intelligence-based approaches can improve the efficacy, responsiveness, and efficiency of diverse systems across multiple disciplines. These systems can adapt and make decisions based on real-time data and analysis thanks to the incorporation of AI technologies.

SERVICES

An arrival spacing technique is also a part of Intelligent Approach services that securely maximize runway capacity in order to boost income, boost operational resilience, improve on-time performance, and lower CO_2 emissions. Its series of functional modules completely integrates with the current air traffic control systems and caters to the specific demands of each airport.

To improve the effectiveness, personalization, and quality of the services provided to clients, intelligence approaches for services integrate AI and data-driven strategies. These strategies seek to enhance client interactions, restructure procedures, and deliver more specialized solutions. Following are some illustrations of intelligence-gathering techniques for services:

Chatbot Customer Support: Businesses utilize chatbots with AI to offer immediate customer service on their websites and messaging services. These chatbots can respond to frequently asked queries, address problems, and direct clients through different procedures.

Personalized Recommendations: Customers receive individualized suggestions based on their tastes and habits via AI-driven recommendation systems. Examples include product recommendations on e-commerce websites and content recommendations on streaming platforms.

Virtual Assistants: NLP is used by virtual assistants like Apple's Siri, Google Assistant, and Amazon's Alexa to assist users with tasks, provide answers to questions, and operate smart devices using voice commands.

Health and Wellness Services: Based on user data from wearables and other sources, health and wellness applications and platforms employ AI to deliver personalized diet and exercise recommendations.

Financial Services: AI algorithms can assess credit risk for lending, offer tailored investing advice, and make suggestions for handling personal finances.

Travel Services: Based on user preferences and past travel information, travel booking companies employ AI to provide tailored trip itineraries, hotel recommendations, and flight possibilities.

Content Personalization: AI is used by content platforms, including social media sites and news websites, to curate content feeds for users, displaying pertinent articles, posts, and updates based on their interests.

Language Translation Services: AI-driven language translation systems can translate written or spoken text in real time, facilitating communication between speakers of various languages.

E-commerce Customer Engagement: AI may examine consumer interactions and behavior to offer personalized marketing messages, promotions, and recommendations to customers.

HR and Recruitment Services: By examining resumes and profiles to match people with job criteria, AI can help with the screening of prospective seekers. Additionally, it helps facilitate interview scheduling and candidate communication.

Education and Learning Platforms: Platforms for education powered by AI can provide students individualized learning routes by customizing the curriculum to suit their strengths and shortcomings.

Subscription Services: Businesses that rely on subscriptions might utilize AI to examine customer usage trends and preferences in order to improve their subscription offerings and pricing structures.

These illustrations highlight how incorporating intelligence into service delivery can enhance client ease, efficiency, and personalization. Utilizing AI, businesses may provide more individualized and flexible services that cater to customer demands and preferences.

EXAMPLES OF INTELLIGENCE APPROACHES FOR OPERATION

Some examples of intelligence approaches applied to operations:

Supply Chain Management: Using AI to examine historical data, market trends, and outside influences to forecast demand patterns is an intelligent approach to supply chain management. This enables businesses to make wise decisions regarding procurement and distribution, optimize inventory levels, and decrease stockouts.

Process Automation: RPA is an intelligence strategy that automates routine processes and workflows using software robots. Data input, invoice processing, and order fulfillment are just examples of jobs that can be automated to free up human workers for more important responsibilities.

Predictive Maintenance: Predictive maintenance, which is used in manufacturing and industrial processes, employs AI to continuously check the health of the equipment. The system can forecast when equipment is likely to break down by examining sensor data and previous maintenance logs, enabling maintenance to be planned in advance.

Anomaly Detection: Operations can check data streams for deviations from expected patterns using AI-powered anomaly detection. For instance, AI in cybersecurity can spot odd network activity that could be a sign of a security breach or cyberattack.

Quality Control and Inspection: Computer vision is used by quality control systems with AI enhancements to inspect products as they are produced. The system ensures that only high-quality products are delivered to clients by detecting flaws, variances, and departures from quality requirements.

Energy Management: By examining data from smart sensors and meters, AI can improve energy use in buildings. On the basis of occupancy patterns and energy demand, this information is used to modify lighting, heating, and cooling systems.

Healthcare Operations: By examining data on patient admissions, discharges, and treatment plans, AI can improve patient flow in hospitals. This assists medical facilities in efficiently allocating resources and reducing patient wait times.

Logistics and Transportation: AI-powered route optimization is a logistics intelligence strategy. The algorithm may choose the most effective routes for delivery vehicles by taking into account variables like traffic, weather, and delivery deadlines.

Financial Operations: Financial data can be analyzed by AI algorithms in real time to spot fraud or irregularities. This aids financial institutions in their efforts to prevent fraudulent transactions and uphold regulatory compliance.

Inventory Optimization: Systems for managing inventories that are AI-driven may analyze historical sales information to forecast product demand. Using this data, inventory levels are optimized, resulting in lower carrying costs and increased product availability.

These illustrations show how intelligence-related approaches can improve diverse operational settings' decision-making accuracy, efficiency, and effectiveness. Organizations can improve their operations' automation, proactive problem-solving, and optimization by utilizing AI technologies.

REFERENCES

1. Moraes, E. B., Kipper, L. M., Hackenhaar Kellermann, A. C., Austria, L., Leivas, P., Moraes, J. A. R., & Witczak, M. (2023). Integration of industry 4.0 technologies with education 4.0: Advantages for improvements in learning. *Interactive Technology and Smart Education*, 20(2), 271–287.
2. Lemstra, M. A. M. S., & de Mesquita, M. A. (2023). Industry 4.0: A tertiary literature review. *Technological Forecasting and Social Change*, 186, 122204.
3. Pozzi, R., Rossi, T., & Secchi, R. (2023). Industry 4.0 technologies: Critical success factors for implementation and improvements in manufacturing companies. *Production Planning & Control*, 34(2), 139–158.
4. Antony, J., Sony, M., & McDermott, O. (2023). Conceptualizing industry 4.0 readiness model dimensions: An exploratory sequential mixed-method study. *The TQM Journal*, 35(2), 577–596.
5. Ding, B., Ferras Hernandez, X., & Agell Jane, N. (2023). Combining lean and agile manufacturing competitive advantages through industry 4.0 technologies: An integrative approach. *Production planning & control*, 34(5), 442–458.
6. Raja Santhi, A., & Muthuswamy, P. (2023). Industry 5.0 or industry 4.0 S? Introduction to industry 4.0 and a peek into the prospective industry 5.0 technologies. *International Journal on Interactive Design and Manufacturing (IJIDeM)*, 17(2), 947–979.
7. Lei, Z., Cai, S., Cui, L., Wu, L., & Liu, Y. (2023). How do different industry 4.0 technologies support certain circular economy practices? *Industrial Management & Data Systems*, 123(4), 1220–1251.
8. Fernando, Y., Tseng, M. L., Wahyuni-Td, I. S., de Sousa Jabbour, A. B. L., Chiappetta Jabbour, C. J., & Foropon, C. (2023). Cyber supply chain risk management and performance in industry 4.0 era: Information system security practices in Malaysia. *Journal of Industrial and Production Engineering*, 40(2), 102–116.
9. Kamble, S. S., & Gunasekaran, A. (2023). Analysing the role of industry 4.0 technologies and circular economy practices in improving sustainable performance in Indian manufacturing organisations. *Production Planning & Control*, 34(10), 887–901.

10. Bianco, D., Bueno, A., Godinho Filho, M., Latan, H., Ganga, G. M. D., Frank, A. G., & Jabbour, C. J. C. (2023). The role of industry 4.0 in developing resilience for manufacturing companies during COVID-19. *International Journal of Production Economics*, 256, 108728.
11. Biswas, B., & Sanyal, M. K. (2019, January). Soft intelligence approaches for selecting products in online market. In *2019 9th International Conference on Cloud Computing, Data Science & Engineering (Confluence)* (pp. 432–437), Uttar Pradesh, India IEEE.

Chapter 8

Parkinson's disease detection using machine learning models

Abhinav Dahiya, Kamaldeep Joshi, and Ravi Saini

INTRODUCTION

Parkinson's disease (PD) is an extremely complex disorder, and there is no accurate scale to estimate its severity. PD is a neurodegenerative ailment that affects motor activities due to a drop in dopamine levels in the brain; consequently, physical repercussions are detected in the body [1]. This neurodegeneration results in a variety of symptoms, such as coordination difficulties, bradykinesia, voice alterations, and rigidity [2]. The lack of neuronal growth capability is the primary cause of PD. As a person matures, neurons begin to die out and are irreplaceable. Dopamine, a chemical fluid produced by neurons, is entirely responsible for movement in the body and signal transmission between neurons. As dopamine levels begin to decline with age, the neurological condition begins to slow down, as a result of the brain's multiple communication mechanisms [3]. Since these effects come on extremely gradually, they are typically not noticeable until the patient's condition has significantly deteriorated. Voice impairment, loss of balance, slow motions, unstable posture, stiffness, sleepiness, facial masking, and other symptoms are some of the things that might occur as a result of this condition.

Disability and death rates resulting from Parkinson's disease (PD) are rising faster than those resulting from any other neurological ailment on a global scale. Over the past quarter of a century, the incidence of PD has more than doubled. According to a report published by the WHO in June 2022 (the most recent statistics from around the world in 2019), there are over 8.5 million people living with PD. PD was responsible for 5.8 million disability-adjusted life years in 2019, representing an increase of 81% since the year 2000. Additionally, it was responsible for 329,000 deaths, representing an increase of over 100% since the year 2000. Therefore, there is a need for diagnostic techniques that have a higher level of sensitivity for the detection of PD because, as the disease advances, additional symptoms appear, which makes it more difficult to treat PD [4].

Loss of intensity, the monotony of pitch and loudness, diminished emphasis, inappropriate silences, short rushes of speech, variable tempo, incorrect consonant articulation, and a harsh and breathy voice are

 DOI: 10.1201/9781003479031-8

the primary characteristics of PD speech (dysphonia). Due to the fact that collecting voice data is not invasive and can be easily done with mobile devices, it seems promising for a possible detection tool to be able to detect a spectrum of voice-related illnesses. Because the initial symptoms of PD are often modest, it can be challenging to diagnose the condition early on. Because of the delays in diagnosis, there is a tremendous strain placed on patients as well as the whole healthcare system [5]. Researchers have been motivated to develop screening techniques that rely on automated algorithms to discriminate healthy controls from persons who have PD due to the difficulties of making an early diagnosis of the condition. The current study is an encouraging initial step toward the long-term objective of offering a decision assistance algorithm for doctors to use when screening patients for PD [6]. In this study, we use many different machine learning models on the Kaggle "Parkinson Disease Detection" Voice dataset in order to differentiate between patients with PD and healthy controls. We began by putting into practice a number of Machine Learning methods, which included Logistic Regression, KNN, Support Vector Machine, Random Forest, Decision Tree, Gaussian Naïve Bayes, Adaptive Boosting, and CatBoost.

The remaining parts have the following organizational structure: The information on the dataset is presented in the next section, which is then followed by the third section, which provides in-depth information about the dataset that was used. Data pre-processing is covered in the fourth section. In the fifth section, the proposed model is discussed. The sixth section explains the obtained results, followed by the seventh section, which discusses the conclusion of the proposed work. The final section discusses the future scope of the proposed work.

DATASET

We utilize the "Parkinson's Disease Detection" dataset supplied by Kaggle [7] in this research. In this study, we classified PD using a Kaggle notebook. Table 8.1 provides detailed information on the characteristics of the dataset. The dataset contains 195 patient records with no missing information (147 for PD patients and 48 for healthy patients).

EXPLORATORY DATA ANALYSIS

We conducted a comprehensive exploratory data analysis on the aforementioned characteristics. The name is being eliminated because it has no relevance to the research. As a result, we gained substantial insight into the impact of these factors on cardiovascular disease. Figure 8.1 shows a heat map for the correlation between features.

Table 8.1   Detailed Information of Features

Features	Description
"Name"	"Patient name and recording number in ASCII format" [7]
"MDVP:Fo (Hz)"	"Average vocal fundamental frequency" [7]
"MDVP:Fhi (Hz)"	"Maximum vocal fundamental frequency" [7]
"MDVP:Flo(Hz)"	"Minimum vocal fundamental frequency" [7]
"MDVP:Jitter(%), MDVP:Jitter(Abs), MDVP:RAP, MDVP:PPQ, Jitter:DDP"	"Five measures of fundamental frequency variation" [7]
"MDVP:Shimmer, MDVP:Shimmer(dB), Shimmer:APQ3, Shimmer:APQ5, MDVP:APQ, Shimmer:DDA"	"Six measures of variation in amplitude" [7]
"NHR,HNR"	"Two measures of ratio of noise to tonal components in the voice" [7]
"RPDE,D2"	"Two nonlinear dynamical complexity measures" [7]
"DFA"	"Signal fractal scaling exponent" [7]
"Spread1, spread2, PPE"	"Three nonlinear fundamental frequency variation measurements" [7]
"Status"	"The subject's health status: one for Parkinson's and zero for good health" [7]

A. MDVP:Fo (Hz)

The value of the correlation between this feature and our target variable is −0.38, which is a negative number and indicates that this feature is only somewhat relevant to this study. As a result, we have retained this functionality for use in future studies.

B. MDVP:FHI (Hz)

The value of the correlation between this feature and our targeted variable is −0.17, which indicates that there is a negative correlation between the two and that this feature is slightly less relevant for this study. As a result, we are going to remove this functionality in order to do more study. It has a low correlation value; hence there is no longer a need for it.

C. MDVP:FIo (Hz)

The value of this feature's correlation with the variable that we are trying to influence is −0.38. Again, because the correlation value is rather high for this study, we decided to save this variable or feature for use in our future research.

D. MDVP:JITTER (%)

The correlation value between this characteristic and our objective variable is 0.28, which indicates that it is positively correlated and meaningful to some extent. Despite the fact that this correlation value is not particularly high, we have decided to keep this feature for the time being in order to conduct further research on it.

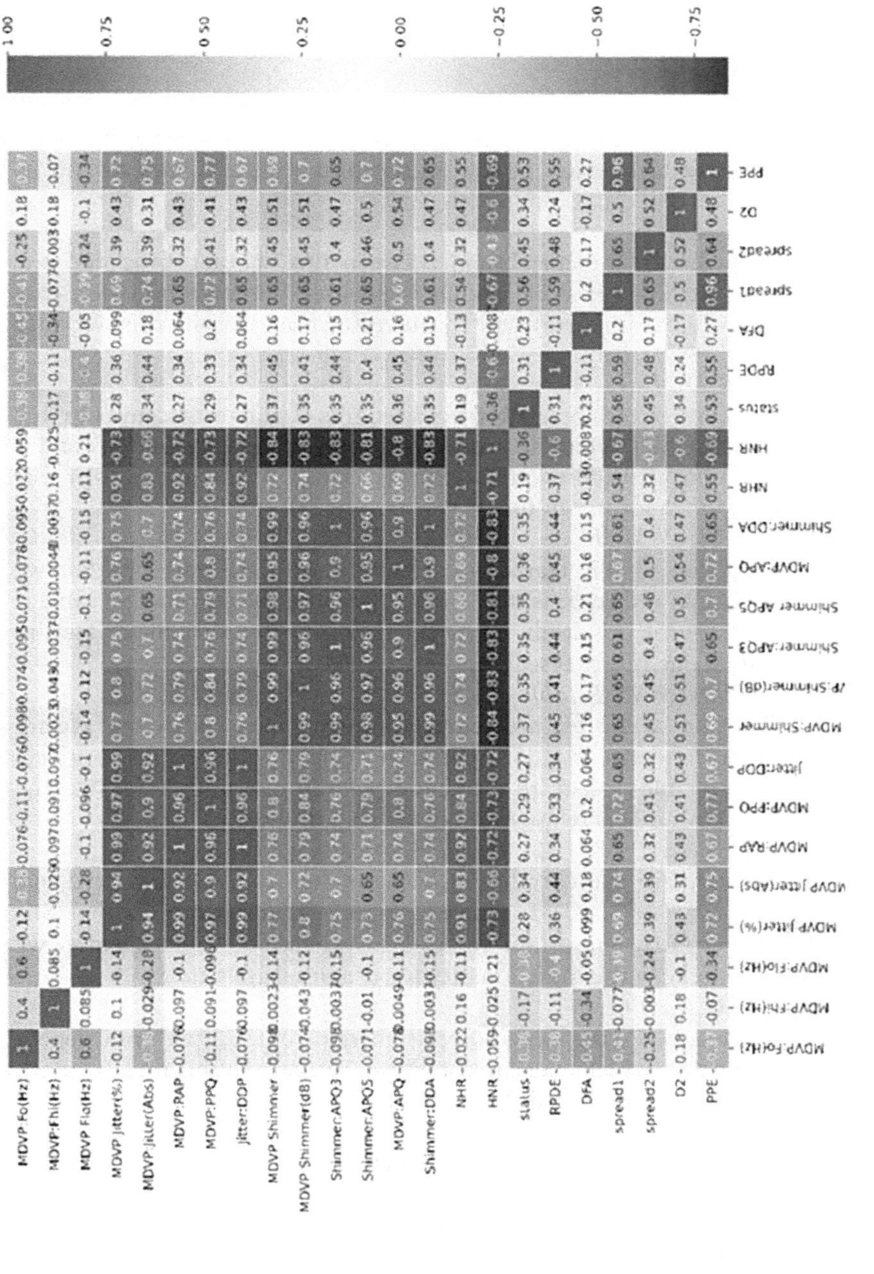

Figure 8.1 Heat map for feature's correlation.

E. *MDVP:JITTER(ABS)*

The correlation value between this property and our primary variable is 0.34, which indicates a good relationship and is extremely pertinent to our investigation. As a result, we decided to keep it for our research.

F. *MDVP:RAP*

The value of this feature's correlation with the variable that we are trying to influence is 0.27. Once more, we have preserved this characteristic or feature for the sake of our future research and additional studies.

G. *MDVP:PPQ*

The value of the correlation between this characteristic and our primary variable is 0.29, which is positive but may be considered marginally less important for this investigation. Despite this, we have retained this feature for use in future studies.

H. *JITTER:DDP*

The value of the correlation between this characteristic and our key variable is 0.27, which is positive but only somewhat important for the purpose of this investigation. Nevertheless, we maintained this characteristic for use in future research.

I. *MDVP:SHIMMER*

The value of this feature's correlation with the variable that we are trying to influence is 0.37. Again, because the correlation value is rather high for this study, we decided to save this variable or feature for use in our future research.

J. *MDVP:SHIMMER (db)*

The value of this feature's correlation with the variable that we are trying to influence is 0.35. Again, we decided to keep this positively linked variable or feature for use in our future research because the correlation value is relatively high for this study.

K. *SHIMMER:APQ3*

The value of the correlation between this characteristic and our primary variable is 0.35, which is favorable and indicates that this feature is fairly important for this investigation. Consequently, we decided to keep this functionality for use in future studies.

L. *SHIMMER:APQ5*

This characteristic has a correlation value of 0.35 with our dependent variable, which is positive and very relevant to this investigation. We therefore maintained this characteristic for further research.

M. *SHIMMER:DDA*

The correlation value between this characteristic and our dependent variable is 0.35. Again, we retained this positively linked variable or characteristic for future research, as the correlation value for this study is fairly high.

N. *MDVP:APQ*

The value of the correlation between this feature and our target variable is 0.36. Again, we kept this positively correlated variable or feature for future research because the correlation value is quite high for this research.

O. *NHR*

The value of the correlation between this characteristic and our primary variable is 0.19, which is positive but only somewhat important for the purpose of this investigation. As a result, we are going to remove this functionality in order to do more study. It has a low correlation value; hence there is no longer a need for it.

P. *HNR*

The value of the correlation between this characteristic and our primary variable is –0.36, which is a negative correlation that is highly significant for this investigation. Consequently, we decided to keep this functionality for use in future studies.

Q. *RPDE*

The correlation value between this characteristic and our dependent variable is 0.31. Again, we retained this positively linked variable or characteristic for future research, as the correlation value for this study is fairly high.

R. *DFA*

The correlation value between this characteristic and our dependent variable is 0.23. This is favorable but somewhat less important to this investigation. We therefore eliminate this attribute for further study. It is being eliminated since its correlation value is low.

S. *SPREAD1*

The correlation value between this characteristic and our dependent variable is 0.56, which is positive and the highest relevance for this study among all characteristics. We therefore maintained this characteristic for further research.

T. *SPREAD2*

This characteristic has a correlation value of 0.45 with our dependent variable, which is positive and very relevant to this investigation. We therefore maintained this characteristic for further research.

U. *D2*

The correlation value between this characteristic and our dependent variable is 0.34. Again, we retained this positively linked variable or characteristic for future research, as the correlation value for this study is fairly high.

V. *PPE*

This characteristic has a correlation value of 0.53 with our dependent variable, which is positive and very relevant to this investigation. We therefore maintained this characteristic for further research.

DATA PRE-PROCESSING

After finishing an in-depth exploratory data analysis, the next step is to begin the pre-processing of the data. The handling of outliers comes first, followed by scaling the data. When dealing with outliers, the initial step is to determine the skew values [8]. Skewed values for each characteristic are displayed in Figure 8.2.

The dataset was altered to fix the skewed values that were found to be above the threshold. After making all of the necessary adjustments, the skewed value of each characteristic has been brought down to a level that is below the threshold value. After the outliers have been handled, the MinMaxScaler [9] is used to scale all of the features. The MinMaxScaler is a specific kind of scaler that adjusts the minimum and maximum values so that they are, respectively, 0 and 1, while the Standard Scalar [10] adjusts all of the values between the minimum and maximum thresholds such that they fit inside the specified range of values from minimum to maximum. As all features are numerical, we don't need to perform any type of encoding.

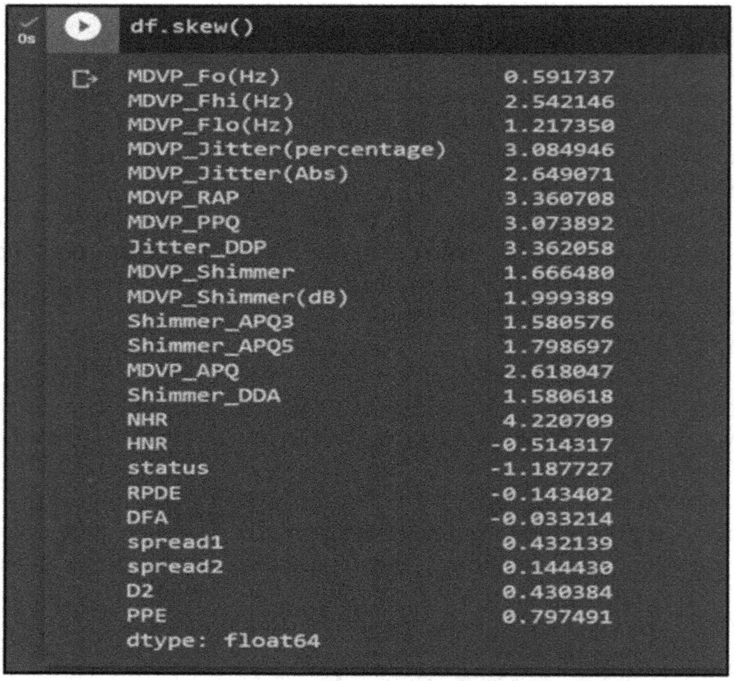

```
df.skew()

MDVP_Fo(Hz)                 0.591737
MDVP_Fhi(Hz)                2.542146
MDVP_Flo(Hz)                1.217350
MDVP_Jitter(percentage)     3.084946
MDVP_Jitter(Abs)            2.649071
MDVP_RAP                    3.360708
MDVP_PPQ                    3.073892
Jitter_DDP                  3.362058
MDVP_Shimmer                1.666480
MDVP_Shimmer(dB)            1.999389
Shimmer_APQ3                1.580576
Shimmer_APQ5                1.798697
MDVP_APQ                    2.618047
Shimmer_DDA                 1.580618
NHR                         4.220709
HNR                        -0.514317
status                     -1.187727
RPDE                       -0.143402
DFA                        -0.033214
spread1                     0.432139
spread2                     0.144430
D2                          0.430384
PPE                         0.797491
dtype: float64
```

Figure 8.2 Skew values of features.

These metrics were used to evaluate the efficacy of these distinct classification strategies:

Accuracy [11]: Accuracy is the proportion of correct predictions made by a model or the number of accurate forecasts relative to the total number of predictions.

$$\text{Accuracy} = \frac{TN + TP}{TN + TP + FP + FN}$$

Recall [11]: Recall is the proportion of true positives that are correctly detected across all predictions.

$$\text{Recall} = \frac{TP}{TP + FN}$$

Precision [11]: It reflects the proportion of actual positives relative to the total number of expected positives.

$$\text{Precision} = \frac{TP}{TP + FP}$$

F1 score [11]: The F1 score represents the harmonic mean of precision and recall.

$$F1 \text{ Score} = \frac{2^* \text{Recall}^* \text{Precision}}{\text{Recall} + \text{Precision}}$$

PROPOSED MODEL

We studied numerous techniques, ranging from simple classifiers such as Logistic Regression to complicated boosting and bagging algorithms such as AdaBoost and CatBoost [12]. Table 8.2 details the classifiers we employed, as well as their accuracy, recall, precision levels, F1 scores, and AUC scores on the test data. Figure 8.3 shows the receiver operating characteristic (ROC) curve graph for all applied classifiers and their ROC values.

RESULTS

This chapter stands out because it provides a complete comparison of numerous PD diagnostic classifiers. In terms of accuracy, precision, recall, F1 score, and AUC score, both CatBoost and K-Nearest Neighbor

Table 8.2 Performance Metrics for Various Applied Machine Learning Algorithms

Classifier	Accuracy	Precision	Recall	F1 Score	AUC Score
Logistic Regression	79.49%	85.29%	90.62%	87.88%	0.90
Decision Tree	89.74%	96.67%	90.62%	93.55%	0.88
SVM	89.74%	96.67%	90.62%	93.55%	0.94
Gaussian Naïve Bayes	61.54%	90.48%	59.38%	71.70%	0.75
KNN	94.87%	96.88%	96.88%	96.88%	0.99
Random Forest	87.18%	93.55%	90.62%	92.06%	0.95
AdaBoost	89.74%	93.75%	93.75%	93.75%	0.91
CatBoost	94.87%	96.88%	96.88%	96.88%	0.99

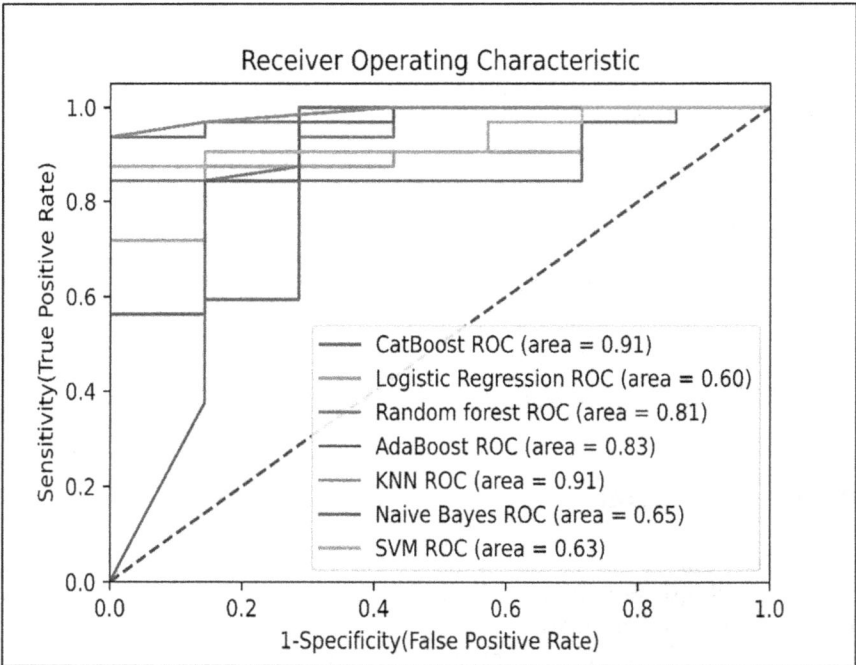

Figure 8.3 ROC curve graph for all classifiers.

outperform other classifiers with an accuracy of 94.87%, the precision of 96.88%, recall of 96.88%, F1 score of 96.88%, and AUC score of 0.99. Again, in terms of ROC curve, CatBoost and K-Nearest Neighbor perform better than the other with area of 91%. Gaussian Naïve Bayes performs worse than the others.

CONCLUSION

In this work, a comparison is made between various different machine learning classifiers for the prediction of PD. The "Parkinson Disease Detection" dataset that is hosted on Kaggle was utilized by our research. The features undergo an in-depth data analysis to determine their effectiveness. We investigated the connections between a wide variety of factors and the roles played by those factors in the development of PD. We utilized the MinMaxScaler technique in order to scale the numerical data. We efficiently locate features that have outliers and then remove such features. We trained numerous machine learning algorithms using these data and then selected the model that was both the most effective and efficient. We came to the conclusion that the models CatBoost and K-Nearest Neighbor are the most efficient ones since they achieved an accuracy of 94.87% and had a recall, precision, and F1 score of 96.88%. The diagnosis of PD in a patient can be determined using this method, which can also be used to avert any future complications.

FUTURE WORK

In the future, additional research could be carried out through the process of further optimizing these models using the hyper-parameter optimization [13] technique. Some of the models have an issue with "over-fitting" the data. In addition, ensemble modeling [14] and neural networking [15] have the potential to produce better results. On account of the larger dataset, additional work could be done.

REFERENCES

1. Wood-Kaczmar, A., Gandhi, S., & Wood, N. W. (2006). Understanding the molecular causes of Parkinson's disease. *Trends in Molecular Medicine*, 12(11), 521–528.
2. Fraser, K. C., Meltzer, J. A., & Rudzicz, F. (2016). Linguistic features identify Alzheimer's disease in narrative speech. *Journal of Alzheimer's Disease*, 49(2), 407–422.
3. Rodriguez, M., Rodriguez-Sabate, C., Morales, I., Sanchez, A., & Sabate, M. (2015). Parkinson's disease as a result of aging. *Aging Cell*, 14(3), 293–308.
4. Armstrong, M. J., & Okun, M. S. (2020). Diagnosis and treatment of Parkinson disease: A review. *JAMA*, 323(6), 548–560.
5. Huse, D. M., Schulman, K., Orsini, L., Castelli-Haley, J., Kennedy, S., & Lenhart, G. (2005). Burden of illness in Parkinson's disease. *Movement Disorders: Official Journal of the Movement Disorder Society*, 20(11), 1449–1454.

6. Tsanas, A., Little, M. A., McSharry, P. E., Spielman, J., & Ramig, L. O. (2012). Novel speech signal processing algorithms for high-accuracy classification of Parkinson's disease. *IEEE Transactions on Biomedical Engineering*, 59(5), 1264–1271.

7. Debasis Samal. (2020). Parkinson Disease Detection Dataset. Retrieved Feb 20, 2023, from https://www.kaggle.com/datasets/debasisdotcom/parkinson-disease-detection

8. Heymann, S., Latapy, M., & Magnien, C. (2012, August). Outskewer: Using skewness to spot outliers in samples and time series. In *2012 IEEE/ACM International Conference on Advances in Social Networks Analysis and Mining* (pp. 527–534), Istanbul, Turkey IEEE.

9. Bisong, E.. (2019). Introduction to Scikit-learn. In *Building Machine Learning and Deep Learning Models on Google Cloud Platform: A Comprehensive Guide for Beginners* (pp. 215–229). Apress, Berkeley, CA.

10. Ferreira, P., Le, D. C., & Zincir-Heywood, N. (2019, October). Exploring feature normalization and temporal information for machine learning based insider threat detection. In *2019 15th International Conference on Network and Service Management (CNSM)* (pp. 1–7), Halifax, Canada IEEE.

11. Yacouby, R., & Axman, D. (2020, November). Probabilistic extension of precision, recall, and F1 score for a more thorough evaluation of classification models. In Proceedings of the First Workshop on Evaluation and Comparison of NLP Systems (pp. 79–91, Association for Computational Linguistics

12. Khoshgoftaar, T. M., Van Hulse, J., & Napolitano, A. (2010). Comparing boosting and bagging techniques with noisy and imbalanced data. *IEEE Transactions on Systems, Man, and Cybernetics-Part A: Systems and Humans*, 41(3), 552–568.

13. Feurer, M., & Hutter, F. (2019). Hyperparameter optimization. In Hutter, F., Kotthoff, L., Vanschoren, J. (eds.), *Automated Machine Learning: Methods, Systems, Challenges* (pp. 3–33). Springer, Cham.

14. Hao, T., Elith, J., Lahoz-Monfort, J. J., & Guillera-Arroita, G. (2020). Testing whether ensemble modelling is advantageous for maximising predictive performance of species distribution models. *Ecography*, 43(4), 549–558.

15. Choi, R. Y., Coyner, A. S., Kalpathy-Cramer, J., Chiang, M. F., & Campbell, J. P. (2020). Introduction to machine learning, neural networks, and deep learning. *Translational Vision Science & Technology*, 9(2), 14–14.

Chapter 9

A comparison between nonlinear mapping and high-resolution image

*Vipul Narayan, Swapnita Srivastava,
Mohammad Faiz, Vimal Kumar,
and Shashank Awasthi*

INTRODUCTION

We trust AI to be the "making of a technique for progress" whose application depends, for every circumstance, on drawing closer to the key figurings
correspondingly as to immense, granular datasets on physical and social
direct. Sorts of progress in neural affiliations and AI henceforth raise the
issue of whether the secret authentic perspectives (i.e., the fundamental
gathered neural affiliation counts) are open. Openings for continued progress in this field—and business applications thereof—are no doubt going to
be on an incredibly major level influenced by terms of approval for pivotal
data. Specifically, if there are loosening up returns to scale or degree in data
procurement (there is more sorting out some way to deal with be had from
the "more important" dataset), it is possible that early or strong people into
a particular application zone may have the decision to make a basic and
trustworthy advantage over potential foes directly through the solicitation
over data rather than through customary got progress or deals side connection impacts. Strong inspirations to keep up data covertly have the additional potential burden that data is not being shared across subject matter
experts; in this manner, lessening the cutoff, thinking about everything, to
get to an amazingly more essential system of data that would rise out of an
open accumulator [1]. To find a point from which we can push toward the
issue of invariant depictions, we should review Mount Castle's estimate of
the one cortical figuring. Is it possible that a singular count is mind-blowing
and versatile enough to sort out some way to see both a catlike and a Beatles
tune? If you are distrustful, consider the immense flexibility the human
psyche shows. It sorts out some way to make invariant depictions subject to
entirely unexpected kinds of data; paying little heed to if individuals are
outwardly weakened, deaf, debilitated, one-looked toward, or puerile, they
sort out some way to make a mental depiction of the world that is savvy and
"right" as in it licenses them to viably investigate and interface with their
ecological components. If we acknowledge that the cortical computation

DOI: 10.1201/9781003479031-9

exists, we ought to assume that it can simply make insignificant doubts about the data it gets. We understand that it should work with visual data on a two-dimensional retina similarly to sound data enlisted as a repeat range in the cochlea. Molecule-savvy development (and that of different affiliations utilizing robotized thinking to move drug divulgence or clinical choice) is still at the beginning stage; in any case, their principal results send an impression of being attractive, and no new courses of action have genuinely come to impart utilizing these new methods of reasoning [2, 3]. In any case, regardless of whether Atom savvy passes on absolutely on its attestation, its improvement is illustrative of the expected endeavor to build up another advancement "playbook", one that usage enormous datasets and learning figures to partake in verifiable suspicion for conventional other-worldly events to energize system persuading mediations. Particle sharp, for instance, is directly sending this way to deal with the exposure and advancement of new pesticides and specialists for controlling yield illnesses. Molecule insightful model depicts two of the paths pushed in man-had thinking that can impact improvement. Regardless, at any rate, the beginnings of man-made consideration are totally in the field of programming, and its fundamental business applications have been in customarily close regions like advanced mechanics. The learning assessments that are directly now being made suggest that man-made discernment may at long last have applications across a wide reach. As indicated by the point of view of the money-related issue of progress (, there is a fundamental package between the issue of equipping improvement motivations to make impels with a quietly little space of use, such robots reason worked for restricted undertakings, versus pushes with a wide—associates may say essentially mind-boggling—zone of use, as might be legitimate for the advances in neural affiliations and AI reliably proposed as "colossal learning." As such, a first mention to be introduced is how much updates in man-made consideration are events of new advances, at any rate rather might be such "all around huge, unexpected turns of events" (later GPTs) that have really been such engaging drivers of expanded length mechanical headway [4]. Second, a couple of places of automated thinking will determinedly set up more moderate or extra stunning obligations to many existing creation measures (punching worries about the potential for goliath occupation clearings); others, as essential learning, hold out the opportunity of not just advantage gains across a wide assortment of regions yet comparatively changes in the legitimate considered the progression relationship inside those spaces. As passed on out and out, by empowering headway across different applications, the "arrangement of a procedure for progress" can have fundamentally more money-related effects than the progress of any single new thing. Here we argue that the new advances in AI and neural relationships through their capacity to improve both the display of end-use impels and the chance of advancement connection, are likely passing all around and influencing development and improvement. As such, the motivations and squares that may shape the unexpected new turn of events and spread of these sorts of progress are a colossal subject for money-related appraisal, and building a point of view on the conditions under which certain potential pioneers can get to these contraptions and utilize them in a significant and ensured way is a focal worry for methodology. This article starts to cripple the possible effect of advances in man-acquired recompense on progress and to see the work that frameworks and establishments may play in giving dazzling motivations to propel, disperse, and dispute around there. We start in Section 9.2 by including the money-related issue of evaluation contraptions, of which fundamental learning applied to R&D issues is a particularly invigorating model. We pivot the exchange between the level of mutilation of the utilization of another appraisal instrument and the piece of evaluation devices, not just in improving the effectiveness of assessment improvement but in making another "playbook" for advancement itself. We by then turn in Section 9.3 to rapidly detach three key innovative headings inside AI—mechanical progress, immense turns of events, and colossal learning. We suggest that these, however much of the time as could be expected, conflated fields will most likely perceive thoroughly astonishing parts later for improvement and unequivocal change. Work in fundamental plans seems to have dropped down and is positively going to have bearably little effect going forward [5]. Besides, exploring those advancements in mechanical progress can likewise get human work in the creation liberated from different things and endeavors; improvement in forefront mechanics moves basically has regularly low potential to change the chance of progress itself. Clearly, epic learning is doubtlessly a region of evaluation that is particularly solid and that can change the advancement correspondence itself. We analyze whether this may doubtlessly be the situation through an appraisal of some quantitative exploratory emphasis on the improvement of various zones of man-made information to the degree strong and express yields of AI specialists as considered (inadequately) by the advancement of papers and licenses from 1990 through 2015. Specifically, we make what we see as the major productive illuminating archive that gets the corpus of target paper and guarantees improvement in man-made sagacious limits, extensively depicted, and isolates these yields into those related to mechanical new development, master developments, and colossal learning. A rich, exact writing analyzing the profitability effects of data innovation highlights the part of the chip as a GPT as a method of understanding the effect of IT on the economy all in all . Different parts of man-made reasoning can positively be perceived as a GPT, and gaining from models, for example, the microchip is probably going to be a helpful establishment for considering both the greatness of their effect on the economy and related arrangement challenges. A second applied system for considering AI is the financial matters of examination apparatuses.

LITERATURE SURVEY

The test introduced by propels in AI is that they seem, by all accounts, to be research instruments that not just can possibly change the technique for advancement itself but in addition have suggestions across an uncommonly wide scope of fields. Truly, advances with these attributes—consider computerized figuring—have had huge and unexpected effects across the economy and society all in all. focuses on the significant effect of IMIs that take the structure not essentially of apparatuses, but rather developments in the way examination is coordinated and directed, like the creation of the college [6]. GPTs that are themselves IMIs (or the other way around) are especially mind-boggling wonders, whose elements are at this point ineffectively comprehended or portrayed. From an arrangement viewpoint, a further significant component of exploration apparatuses is that it very well might be especially hard to suit their advantages. As accentuated, giving suitable motivators to an upstream trend-setter that grows just the principal "stage" of an advancement (e.g., an exploration instrument) can be especially tricky when contracting is defective, and a definitive use of the new items whose improvement is empowered by the upstream development is dubious [7]. Scotchmer and her co-creators underlined a central issue about a multi-stage research measure: when a definitive advancement that makes esteem requires various advances, giving suitable development inc. In his omnibus authentic record of AI research, characterizes AI as "that action gave to making machines wise, and knowledge is that quality that empowers a substance to work fittingly and with premonition in its current circumstance [8]." His record subtleties the commitments of different fields to accomplishments in AI, including yet not restricted to science, semantics, brain research and intellectual sciences, neuroscience, arithmetic, reasoning, and rationale, designing, and software engineering.

METHODOLOGY

Inside the examination areas, a few developments open new roads of request or just improve efficiency "inside the lab." A portion of these advances seem to have extraordinary potential across a wide arrangement of spaces, past their underlying application. As featured by in his exemplary investigations of mixture corn, some new exploration instruments are innovations that do not simply make or improve a particular item—rather they establish another method of making new items with a lot more extensive applications. In Griliches' celebrated development, the disclosure of deceive hybridization "was the innovation of a strategy for imagining." (Hereinafter, "IMI".) Rather than being a method for making a solitary another corn assortment, cross-breed corn addressed a broadly relevant technique for rearing a wide range of new assortments. At the point when applied to the test of

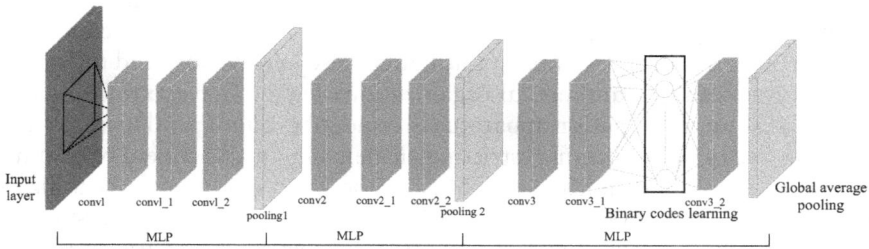

Figure 9.1 Hidden layer of trained neural network.

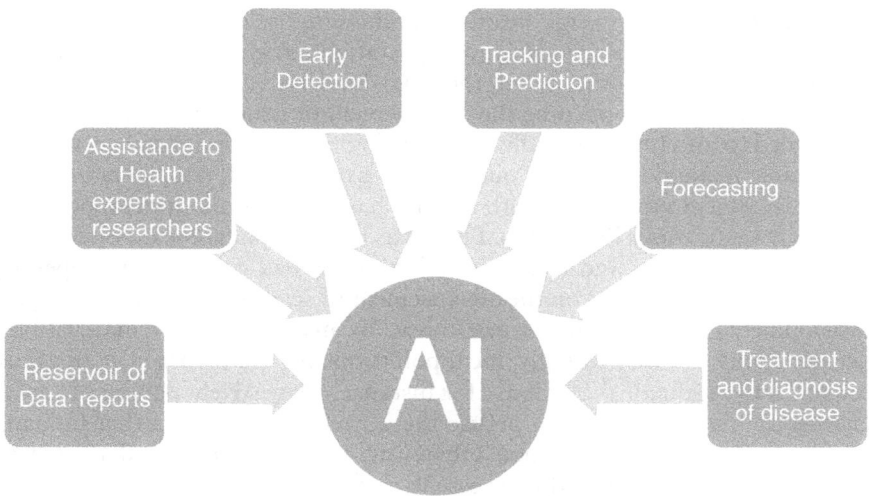

Figure 9.2 Applications of artificial intelligence include in the intro part.

making new assortments upgraded for various areas (and surprisingly more comprehensively, to different harvests) the creation of betray hybridization immensely affected horticultural efficiency (Figures 9.1 and 9.2).

One of the significant bits of knowledge to be acquired from pondering IMIs, thusly, is that the monetary effect of certain kinds of examination apparatuses is not restricted to their capacity to lessen the expenses of explicit development exercises—maybe much more importantly they empower another way to deal with advancement itself, by changing the "playbook" for development in the areas where the new device is applied. For instance, before the methodical comprehension of the force of "half breed energy," an essential concentration in horticulture had been improved procedures for self-treatment (i.e., considering an ever-increasing number of particular regular varietals over the long haul). When the standards overseeing hybridization (i.e., heterosis) were arranged and the presentation benefits of cross-breed power illustrated, the strategies and applied methodology for agrarian advancement were moved, introducing an extensive stretch of methodical development utilizing these new devices and information. Advances in AI and neural organizations seem to have extraordinary potential as an examination apparatus in issues of grouping and forecasting.

These are both significant restricting elements in an assortment of examination undertakings, and, as exemplified by the Atom wise model, the utilization of "learning" ways to deal with AI holds out the possibility of drastically lower costs and improved execution in R&D projects where these are huge difficulties. In any case, likewise, with half and half corn, AI-based learning might be more conveniently comprehended as an IMI than as a barely restricted answer for a particular issue. On the one hand, AI-based learning might have the option to significantly "mechanize disclosure" across numerous spaces where arrangement and expectation assignments assume a significant part. On the other hand, they may likewise "extend the playbook" which is the feeling of opening the arrangement of issues that can be plausibly tended to and fundamentally changing logical and specialized networks' reasonable methodologies and outlining of issues. The development of optical focal points in the seventeenth century had a significant direct monetary effect on applications like scenes. Be that as it may, optical focal points such as magnifying lenses and telescopes additionally had colossal and dependable roundabout impacts on the advancement of science, innovative change, development, and government assistance. By making little or inaccessible articles noticeable interestingly, focal points opened altogether new areas of request and mechanical freedom., for instance, suggestively portray AI as a chance to "figure out how to peruse the genome" in manners that human discernment and insight cannot. Obviously, many exploration devices are neither IMIs nor GPTs, and their essential effect is to lessen the expense or upgrade the nature of a current development measure. For instance, in the drug business, new sorts of materials guarantee to upgrade the effectiveness of explicit examination measures. Other examination devices can certainly be considered as IMIs; however, they are generally restricted in application. For instance, the improvement of hereditarily designed examination mice (like the Oncourse) is an IMI that significantly affects the lead and "playbook" of biomedical exploration, however, has no undeniable importance to advancement in regions like data innovation, energy, or aviation.

RESULT

It is obviously conceivable that this field will see advancements later; however most would agree that, while representative frameworks keep on being a territory of scholarly examination, they have not been fundamental to the business use of AI. Nor is it at the core of the new detailed advances in AI that are related to the region of AI and forecast. A second compelling direction in AI has been extensively near mechanical technology.

While the idea of "robots" as machines that can perform human undertakings goes back in any event to the 1940s, the field of mechanical technology started to seriously thrive from the 1980s onward through a mix of advances in mathematically controlled machine devices and the improvement of more versatile yet at the same time decided to put together mechanical technology that depends with respect to the dynamic detection of a known climate. Maybe the most financially significant use of AI to date has been here, with the huge scope of arrangement of "modern robots" in assembling applications. Notwithstanding the way that head in nature (and clearly defective given that fundamental pieces of appraisal action in man-made consideration may not be unmistakable utilizing these standard development assessments), we discover striking proof for a brisk and epic move in the application orientation of learning-focused dispersals, especially after 2009. The state of this move is useful since it agrees with vivacious authentication about the phenomenally solid execution of accumulated "enormous learning" complex neural relationship in a degree of tries including PC vision and other figure assignments. Solid check (not pronounced here) considering the reference advisors for producers, for example, Geoffrey Hinton who is driving figure in fundamental learning, proposes a striking rate extension of work over two or three years that makes modestly barely any algorithmic disclosures identified with complex neural affiliations.

CONCLUSION

Also, obviously, paying little mind to their specific methodology, man-made consciousness research has been joined from the start by its commitment with Although frequently gathered, the scholarly history of AI as a logical and specialized field is helpfully educated by recognizing three interrelated yet separate territories: advanced mechanics, neural organizations, and emblematic frameworks. Maybe the best line of examination in the early long periods of AI—tracing all the way back to the 1960s—falls under the expansive heading of representative frameworks [9,10]. Albeit early pioneers, for example, Turing had underlined the significance of encouraging a machine as one would a kid (i.e., accentuating AI as a learning cycle), the "image handling speculation" was started on the endeavor to duplicate the coherent progression of human dynamic through preparing images. Early endeavors to start up this methodology yielded striking accomplishments in show projects, for example, the capacity of a PC to explore components of a chess game (or other prepackaged games) or participate in generally straightforward discussions with people by observing explicit heuristics and rules implanted into a program. In any case, while research dependent on the idea of a "general issue solver" has kept on being a territory of huge scholarly interest, and there have been occasional blasts of interest in the utilization of such ways to deal with the human dynamic (e.g., with regard to beginning phase master frameworks to control clinical determination), the representative frameworks approach has been vigorously condemned for its powerlessness to genuinely affect certifiable cycles in an adaptable manner.

REFERENCES

[1] M. Bedny et al.: "Language Processing in the Occipital Cortex of Congenitally Blind Adults." *Proceedings of the National Academy of Sciences* 108, no. 11 (March 15, 2011) 4429–4434.

[2] N. Bostrom: *Superintelligence - Paths, Dangers, Strategies* (Oxford University Press, 2014).

[3] E. Brynjolfsson, A. McAfee: *The Second Machine Age* (Norton Paperback, 2016).

[4] J. J. DiCarlo, D. Zoccolan, N. C. Rust: "How Does the Brain Solve Visual Object Recognition?" *Neuron* 73, no. 3 (February 9, 2012) 415–434. doi:10.1016/j.neuron.2012.01.010.

[5] D. E. Feldmann: "Synaptic Mechanisms for Plasticity in Neocortex." *Annual Review of Neuroscience* 32 (2009) 33–55.

[6] B. Fritzke: "A growing neural gas network learns topologies." In: Tesauro, G.; Touretzky, D. S. and Leen, T. K. (Eds.), *Advances in Neural Information Processing Systems 7*, MIT Press, 1995, 625–632.

[7] Future of Life Institute: "An Open Letter - Research Priorities for Robust and Beneficial Artificial Intelligence." https://futureoflife.org/ai-open-letter/?cn-reloaded=1.

[8] J. Hawkins, S. Ahmad: "Why Neurons Have Thousands of Synapses: A Theory of Sequence Memory in Neocortex." *Frontiers in Neural Circuits* 10 (2016) 1–13.

[9] J. Hawkins, S. Blakeslee: *On Intelligence* (Holt Paperbacks, 2005).

[10] J. Hawkins, M. Lewis, S. Purdy, M. Klukas, S. Ahmad: "A Framework for Intelligence and Cortical Function Based on Grid Cells in the Neocortex." *Frontiers in Neural Circuits* 12 (2018) 121.

Chapter 10

Deployment issues in industrial resolution

Abhishek Kumar Shukla and Anil Kumar Dubey

A delicate balance must be struck between technology advancement, operational needs, and safety considerations when deploying software solutions in industrial settings. To negotiate these difficulties and accomplish successful software deployment, careful planning, communication between IT and operations teams, and adherence to best practices are necessary (Table 10.1).

ISSUES IN INDUSTRIAL REVOLUTION 4.0

Industrial Revolution 4.0, also known as Industry 4.0 or the Fourth Industrial Revolution, is the transformation of industrial systems and processes through the integration of digital technology [8], data analytics, automation, and connectivity. Industry 4.0 has a lot of potential advantages, but there are also concerns and obstacles that must be resolved. These are some of the main problems with implementing Industry 4.0 (Table 10.2).

Technological innovation, strategic planning, cooperation, and adaptation must all be used to address these difficulties. Although Industry 4.0 promises to have a transformative impact, businesses must carefully negotiate these difficulties in order to realize the promised improvements in productivity, efficiency, and competitiveness.

Data security and privacy

Data security and privacy [7, 9] are of utmost importance in the context of Industry 4.0, where more connection, data sharing, and digitalization provide new opportunities but also potentially dangerous situations. The deployment of Industry 4.0 must be successful in order to protect sensitive data, preserve system integrity, and guarantee privacy laws are followed. The following are crucial things to keep in mind when tackling data security and privacy issues.

Table 10.1 Typical deployment concern in industrial setting

S. No.	Key features	Issue	Consideration
1	Compatibility and integration	Ensuring that new software is compatible with the systems, gadgets, and communication protocols now in use in industrial settings	To find potential issues and ensure smooth interaction with current infrastructure, perform comprehensive compatibility and integration testing.
2	Downtime and disruption	Minimizing downtime and interference with active industrial processes while deploying software	To lessen the impact on production processes, schedule deployment for times when there are no maintenance windows or during off-peak hours. Use rollback techniques if problems occur.
3	Security and reliability	To avoid cyberattacks, data breaches, and system failures, the new software solution's security and dependability must be guaranteed.	Implement appropriate authentication and authorization procedures, conduct in-depth security testing, and adhere to best practices for industrial cybersecurity.
4	Regulatory compliance	Observing industry-specific rules and guidelines for security, data privacy, and quality control	Make sure that the new software solution conforms with all applicable laws and standards, and record your efforts to do so.
5	Legacy systems	Incorporating new software with existing systems, some of which may use dated technology and have limited interoperability	To enable data exchange and communication between contemporary software and legacy systems, use middleware, adapters, or APIs.
6	Remote locations and connectivity	Deploying software in remote industrial locations with spotty network connectivity	When connectivity is sporadic, make plans for offline functionality and data syncing. Software should be improved for low-bandwidth data transfer efficiency.
7	Training and user adoption	Ensuring successful user uptake of the new software and enough training for industrial personnel to use it	Users should receive thorough training, and you should include them in the software's testing and validation to deal with usability concerns as soon as they arise.
8	Scalability and performance	Making sure the software solution can expand to meet rising demand for industrial operations	Test the software's capacity to manage increasing loads and data quantities by doing load tests. Beginning with scalability in mind, design.
9	Environmental factors	Taking into account environmental elements that may have an impact on software and hardware components, such as temperature, humidity, and physical wear and tear	To endure the challenging environmental conditions frequently encountered in industrial environments, choose ruggedized hardware and design software.
10	Vendor support and maintenance	Ensuring continuing vendor maintenance, updates, and support for the installed software solution	To resolve security vulnerabilities and ensure compatibility, pick credible vendors with a history of offering ongoing support and updates.

Table 10.2 Issue with implementation of Industry 4.0

S. No.	Key feature	Issue	Consideration
I	Data security and privacy	Concerns regarding data security and privacy are raised by Industry 4.0's growing connection and data sharing, particularly in important industrial areas.	To protect sensitive data, take effective cybersecurity measures, encryption, access controls, and data protection compliance.
2	Interoperability and standardization	Interoperability issues may arise from the integration of various technologies and systems from various manufacturers.	To guarantee seamless communication and backward compatibility across devices and systems, create and adopt industry-wide standards and protocols.
3	Skilled workforce shortage	The transition to digital technologies necessitates the development of a competent workforce with expertise in data analytics, IoT, cybersecurity, and other pertinent fields.	To close the skills gap and prepare people for new jobs, fund workforce development initiatives.
4	High initial investment	Industry 4.0 technology adoption frequently necessitates a substantial upfront investment in hardware, software, infrastructure, and training.	To justify the expenditure, create a convincing business case, conduct a ROI analysis, and take into account phased imple-mentations to control expenses.
5	Change management and cultural shift	Resistance from employees and management might arise when switching from traditionally driven procedures to digitally driven ones.	Encourage an innovative culture, be transparent about the advantages of Industry 4.0, and include staff in the transformation.
6	Data overload and analysis	IoT devices and sensors generate a lot of data, which can make it difficult to analyze that data efficiently and derive insights that can be put to use.	Process, analyze, and glean meaningful insights from huge datasets by using data analytics tools and methodologies.
7	Regulatory and legal challenges	Industry 4.0 may present fresh legal and regulatory obstacles pertaining to data ownership, responsibility, and compliance.	To ensure adherence to applicable laws, stay informed about changing legislation and consult with legal professionals.
8	Maintenance and upgrades	It can be challenging to maintain and keep Industry 4.0 systems up-to-date, particularly in settings with older equipment.	To maintain the dependability and life span of digital systems, schedule routine maintenance, updates, and upgrades.
9	Energy consumption and sustainability	Increased digitization might result in increased energy use, which would affect how environmentally sustainable industrial operations are overall.	To reduce your impact on the environment, use energy-efficient technologies and keep an eye on your energy consumption.
10	Rural and small-scale adoption	It may be difficult for smaller businesses and rural locations to access and deploy Industry 4.0 technologies.	Governments, business organizations, and other organizations may offer assistance, resources, and rewards to encourage adoption in various fields.

- **Data Encryption:** Use effective encryption techniques to safeguard data while it is in transit and at rest. This stops illegal access and guarantees that even if data is captured, it will remain unreadable in the absence of the necessary decryption keys.
- **Access Controls and Authentication:** Implement strong access controls to limit access to data to only authorized personnel. Make use of MFA (multi-factor authentication) to guarantee that only authorized users have access to vital systems and data.
- **Data Minimization:** Only gather and keep the data that is required for the aforementioned purposes. The danger of exposure and the possible consequences of a data breach are reduced by reducing the amount of data collected.
- **Secure Communication Protocols:** To guarantee that data sent between devices and systems remains private and untampered with, use secure communication protocols like HTTPS and MQTT with security extensions.
- **Network Segmentation:** Networks can be segmented to separate important systems and sensitive data from less important components. As a result, the effect and possible spread of security breaches are constrained.
- **IoT Device Security:** Use secure boot procedures, firmware updates, and device authentication as security measures to be implemented at the device level. IoT devices that are not secure can be used as attack entry points.
- **Regular Updates and Patch Management:** Update all hardware, software, and devices with the most recent security updates to close known security holes and thwart hacker attempts.
- **Security Audits and Testing:** To find and fix holes in systems and applications, and conduct routine security audits, vulnerability assessments, and penetration tests.
- **Privacy by Design:** Design systems, applications, and processes with privacy considerations are integrated from the beginning. Implement privacy measures that give users access to their data.
- **Employee Training and Awareness:** Inform staff members and other interested parties about social engineering threats, best practices for cybersecurity, and the value of data protection and privacy.
- **Data Breach Response Plan:** Create a thorough data breach response strategy that outlines what you do in the event of a security incident. This plan should contain communication tactics and measures to lessen the effects of a breach.
- **Compliance with Regulations:** Keep up with local and industry-specific data protection and privacy laws (such as CCPA and GDPR). Make sure that your data practices adhere to these laws.

- **Third-Party Vendor Security:** Look into the security procedures used by partners and third-party vendors who have access to your data to ensure they adhere to your standards for privacy and security.
- **Continuous Monitoring:** Utilize intrusion detection and continuous monitoring to quickly detect and address security risks or irregularities.

Industry 4.0 demands a comprehensive strategy to data security and privacy that includes technology, rules, processes, and a strong commitment to protecting sensitive information. Organizations can benefit from digitization while reducing risks to their operations and reputation by taking proactive steps to secure data and respect user privacy.

Example of data security and privacy in Industry 4.0

Some examples illustrating data security and privacy considerations in the context of Industry 4.0:

- **Scenario:** Imagine a smart manufacturing facility that has embraced Industry 4.0 concepts to enhance productivity, boost effectiveness, and enable preventive maintenance. The factory is furnished with a variety of IoT gadgets [6], sensors, and connected systems that gather and process real-time data to guarantee efficient operations. Here is how data privacy and security are handled.
- **Data Encryption:** Secure communication protocols, such as MQTT with TLS/SSL, are used to encrypt all data sent between IoT devices, sensors, and control systems within the plant to prevent unauthorized access.
- **Access Controls and Authentication:** Access to the plant's control systems and data is restricted to individuals with the appropriate credentials. Critical systems require MFA before being accessed.
- **IoT Device Security:** To guard against unwanted tampering or hacking, IoT devices have embedded security features including secure boot procedures and digitally signed firmware.
- **Network Segmentation:** The network of the plant is divided into sections that separate information technology (IT) systems from operational technology (OT) systems, lowering the attack surface and reducing the severity of breaches.
- **Privacy by Design:** Systems for collecting and processing data are made to acquire sensitive data and personally identifying information (PII) as little as possible. Anonymization of data is used when possible.
- **Regular Updates and Patch Management:** To fix known vulnerabilities and security problems, all devices and systems receive routine firmware and security patch upgrades.

- **Security Audits and Testing:** To find vulnerabilities and evaluate the efficacy of security procedures, periodic security audits and penetration tests are carried out.
- **Employee Training and Awareness:** Employees at the plant often undergo training on cybersecurity best practices, such as identifying phishing efforts and safeguarding sensitive information.
- **Vendor Security:** To make sure their goods match the plant's security standards, third-party vendors who offer IoT devices or software solutions go through extensive security examinations.
- **Data Breach Response Plan:** The facility has a detailed data breach response plan that describes what to do in the event of a breach, including alerting the relevant parties and minimizing the effects.
- **Compliance with Regulations:** The facility adjusts its procedures in accordance with industry-specific laws, such as GDPR and ISO 27001 for data protection.
- **Continuous Monitoring:** To enable quick reactions to possible threats, intrusion detection systems continuously scan network data for any indications of anomalous activity.

The smart manufacturing [13, 14, 15, 16] facility ensures that its Industry 4.0 activities are supported by strong data security and privacy standards by putting these safeguards into place. This enables the facility to use digital technology for enhanced operations while protecting sensitive information and upholding the confidence of its stakeholders and consumers.

Interoperability and standardization

Industry 4.0 is dependent on interoperability and standardization because they enable easy communication, teamwork, and integration between various technologies, systems, and devices. While standardization creates common frameworks [1, 5, 9, 21], protocols, and formats that enable consistency and compatibility, interoperability ensures that various components may function together efficiently. Here's a closer look at the significance of standardization and interoperability in Industry 4.0.

Interoperability

- **Collaboration of Heterogeneous Systems:** Systems, devices, and technologies from many vendors must work together in Industry 4.0 to accomplish shared objectives. Interoperability guarantees that these elements can interact, share information, and work together effectively.
- **Optimized Processes:** When multiple components, including sensors, machines, and software applications, can communicate with one another in real time, production processes are optimized, downtime is decreased, and efficiency is increased.

- **Flexibility and Scalability:** Manufacturers are able to adopt new technologies and devices without affecting current operations thanks to interoperable systems' increased flexibility and ability to react to changes.
- **Data Sharing:** Interoperable systems allow for the frictionless interchange of data throughout diverse organizational departments, improving decision-making and process visibility.

Standardization

- **Consistency:** Common practices, protocols, and interfaces are established through standardization, guaranteeing uniformity in how various components interact and communicate.
- **Reduced Complexity:** Standardized interfaces and protocols make integration tasks easier by giving developers a common framework to follow, which lowers complexity and speeds up development.
- **Vendor Neutrality:** Standards encourage solutions that are independent of the vendor, enabling manufacturers to select the top components from many vendors while ensuring compatibility.
- **Interchangeability:** Standardization enables simple system or component replacement without the need for extensive alterations or unique integrations.
- **Easier Maintenance:** Because modifications to one component are less likely to have an adverse effect on the ecosystem as a whole, standardized systems are simpler to maintain and upgrade.

Challenges and considerations

- **Legacy Systems:** It might be difficult to integrate older systems with more modern, compatible technologies because of disparities in technology and communication standards.
- **Vendor Adoption:** To ensure seamless interoperability, vendors must embrace and follow industry standards.
- **Complex Ecosystems:** Achieving complete interoperability can be difficult due to the complexity of modern industrial ecosystems, which include a variety of technologies, devices, and protocols.
- **Evolution of Standards:** Industry 4.0 is a developing sector; therefore, norms could alter over time. To retain compatibility, manufacturers must keep up with changing standards.
- **Balancing Customization:** Despite the advantages of standardization, producers need some degree of customization to fulfill certain operating needs.

Benefits

- **Efficiency:** Standardization and interoperability result in streamlined processes, fewer manual interventions, and effective data interchange.

- **Innovation:** By allowing for the quick integration of new technology and solutions, a standardized ecosystem promotes innovation.
- **Flexibility:** By adding or replacing components, manufacturers can quickly respond to shifting business needs without disturbing the system as a whole.
- **Collaboration:** Industry-wide standards let firms collaborate and share their best practices and solutions.
- **Time and Cost Savings:** Cost and time benefits result from less integration efforts, quicker development cycles, and fewer disruptions.

Realizing Industry 4.0's full potential requires interoperability and standardization. These principles establish the groundwork for a more connected and effective industrial landscape by encouraging collaboration, minimizing complexity, and facilitating seamless communication.

Example: smart factory automation

Imagine a smart factory [10, 11, 12, 17, 18, 19] where a variety of equipment, sensors, and software programs collaborate to automate production procedures. Smooth operations and effective manufacturing depend heavily on interoperability and standardization.

Interoperability

- **Collaboration of Machines:** To create a finished product, various machine types—including CNC machines, robotic arms, and conveyor systems—need to work together. These robots can interact, share information, and plan their activities because of interoperability.
- **Data Exchange with Sensors:** The factory is equipped with sensors that measure pressure, humidity, and other variables. These sensors can transfer data in real time to control systems that can modify manufacturing processes based on the information thanks to interoperability.
- **Integration of Software Applications:** Data communication between software programs is necessary for systems like factory execution systems (MES), enterprise resource planning (ERP) systems, and quality control software. By ensuring information flows between various apps, interoperability enables effective decision-making and process optimization.

Standardization

- **Common Communication Protocols:** Standardized communication protocols, like OPC UA (Unified Architecture [3]), guarantee that hardware, software, and sensors all speak the same language when exchanging data. This decreases complexity and does away with the requirement for custom integration work.

- **Data Formats:** To facilitate easy interpretation and usage, standardized data formats make sure that information transferred between various components is organized consistently.
- **Plug-and-Play Compatibility:** Machines and equipment from many manufacturers can be "plug-and-play" compatible thanks to standardized interfaces. For instance, utilizing standardized interfaces, a robotic arm from one manufacturer can be easily linked with a conveyor system from another vendor.

Benefits

- **Efficient Production:** Interoperability enables equipment to cooperate smoothly, removing bottlenecks and improving manufacturing procedures.
- **Real-Time Insights:** Control systems and software applications can easily access sensor data, enabling real-time monitoring and modifications.
- **Quality Control:** Automated quality inspections are made possible by interoperability and standardization, guaranteeing that goods adhere to predetermined norms.
- **Flexibility:** Standardized interfaces make it simple for the factory to replace or add new machines without requiring extensive custom integrations.
- **Data-Driven Decisions:** Software programs can handle and analyze data more easily thanks to standardized data formats, enabling data-driven decision-making.

Challenges and considerations

- **Legacy Systems:** Due to differences in technology and protocols, integrating legacy equipment with contemporary, compatible systems can be difficult.
- **Vendor Adoption:** It is essential to make sure that equipment and devices from various suppliers implement standardized protocols and interfaces.
- **Maintenance:** Standardized systems are simpler to upgrade and maintain, but it's crucial to make sure that standards are consistently followed.
- **Evolution of Standards:** As technology develops, manufacturers must keep up with changing standards to ensure compatibility.

The smart factory may achieve greater levels of automation, efficiency, and collaboration across its machines, systems, and software applications by using interoperable and standardized solutions. This results in enhanced performance overall, decreased downtime, and optimized production.

Skilled workforce shortage

As more sectors implement cutting-edge technology and automation, they face a huge challenge: a skilled labor shortage in the context of Industry 4.0, encompassing IoT, data analytics, artificial intelligence, and automation, highlights the demand for specialized skills that are frequently in short supply. Here's a summary of the problems and solutions that businesses can use to resolve them.

Issue

- **Skills Mismatch:** The skills needed by Industry 4.0 technology and the skills already in the workforce are incompatible. Working with cutting-edge digital technologies may not automatically translate traditional industrial abilities.
- **Rapid Technological Advancements:** Industry 4.0 technologies are developing quickly, making it difficult for the workforce to stay abreast of the most recent advancements and abilities.
- **Multi-disciplinary Skills:** Industry 4.0 demands a combination of abilities from several fields, such as engineering, data science, cybersecurity, and IT, which can be challenging to locate in a single person.
- **Lack of Training:** Due to the slow adoption of Industry 4.0 by many educational institutions, there is a dearth of graduates with the necessary skills.
- **Competition:** Increased competition among industries for attracting and maintaining top talent is a result of the high demand for qualified individuals in the area.

Addressing the shortage

- **Training and Education:** Develop training programs and curricula with educational institutions that are focused on Industry 4.0 and provide students the skills they need.
- **Internal Training Programs:** Create internal training programs to upskill current staff and close the skills gap. This may entail seminars, online classes, and practical instruction.
- **Upskilling and Reskilling:** Provide opportunity for employees who have the potential to upgrade their skills or move into positions requiring knowledge of Industry 4.0.
- **Partnerships with Tech Providers:** Offer training and certification courses on the platforms and tools of technology vendors.
- **Apprenticeships and Internships:** Create apprenticeship and internship programs to give students and early-career professionals practical experience.

- **Cross-Functional Teams:** Form cross-functional teams with people with a variety of expertise to address complicated industry issues.
- **Professional Development:** By giving workers resources, time, and rewards for skill improvement, you may promote continual learning and professional development.
- **Industry-Academia Partnerships:** Participate in joint initiatives that bridge the gap between academic knowledge and practical skills by working with universities and research organizations.
- **Remote Work and Global Talent:** To get beyond geographic restrictions in locating qualified specialists, take into account remote work choices and utilize global talent pools.
- **Company Culture:** To draw and keep talent interested in Industry 4.0 and promote a culture of learning, innovation, and adaptability.
- **Diversity and Inclusion:** Encourage diversity and inclusion programs to draw more people with diverse backgrounds and viewpoints.

Industry 4.0 requires a multifaceted strategy to address the labor crisis, including cooperation between business, academia, and government, as well as a dedication to continual learning and growth within enterprises.

Example: smart manufacturing facility

Consider a manufacturing plant that wants to deploy Industry 4.0 technology to improve the effectiveness of its production operations and lower downtime. To accomplish these objectives, the facility intends to use automation, data analytics, and IoT devices. However, there are substantial obstacles due to a lack of skilled workers with the necessary knowledge of Industry 4.0:

Scenario

IoT Implementation: The plant plans to add Internet of Things (IoT) sensors to its production equipment to track machine performance, temperature, and energy consumption [26] in real time. However, they are short on personnel with expertise in data integration, analytics, and sensor deployment.

Challenges

- **Skills Gap:** The current workforce lacks the skills necessary to set up, configure, manage, and evaluate the data produced by IoT devices.
- **Data Analytics:** Finding personnel who can efficiently analyze the data gathered from IoT sensors to spot trends, abnormalities, and improvement opportunities is a challenge for the facility.
- **Integration:** Data integration, API usage, and software development expertise are necessary for integrating IoT data into existing systems.

Addressing the shortage

- **Training Programs:** To create training programs with an emphasis on IoT deployment, data analytics, and integration, the facility collaborates with nearby universities and technical schools. Students will get practical knowledge and Industry 4.0-related skills as a result of this.
- **Internal Training:** The facility finds promising workers and gives them specific training in data analytics and IoT technology. This method fills the skills gap among current employees.
- **Collaboration with Experts:** For employee workshops and seminars, the facility works with technology vendors and IoT specialists. Employee exposure to industry thought leaders and practical knowledge aids learning.
- **Outsourcing:** The facility may collaborate with outside consultants or service providers who are qualified in IoT deployment and data analytics in order to address urgent demands.
- **Talent Acquisition:** To fill key positions, the facility aggressively seeks for candidates with experience in data analytics, IoT, and automation. It might think about providing lucrative compensation packages to entice outstanding talent.
- **Cross-Training:** To develop their IoT and data analytics skills, employees from several departments receive cross-training. This enables flexibility and knowledge sharing between teams.
- **Collaborative Projects:** To have access to specialized skills and information, the facility collaborates on projects with universities, research institutions, and technology firms.

The manufacturing plant may successfully deploy IoT technology and use data analytics to enhance its production processes, save costs, and boost overall efficiency by tackling the skilled manpower shortage through strategic training, collaboration, and talent acquisition.

High initial investment

Organizations looking to adopt cutting-edge digital solutions frequently struggle with the substantial initial expenditure necessary for integrating Industry 4.0 technologies. Although Industry 4.0 promises considerable long-term advantages like improved production, efficiency, and innovation, the initial expenses can be high. Here is a look at the problem and some possible solutions:

Issue

- **Technology Costs:** It can be expensive to purchase sensors, IoT devices, automation systems, data analytics platforms, and other Industry 4.0 technologies, particularly when outfitting a full facility or updating existing infrastructure.

- **Infrastructure:** To accommodate new technologies, old infrastructure may need to be upgraded, which could cost money for network improvements, cybersecurity measures, and compatibility improvements.
- **Training:** The personnel must be properly taught to utilize and maintain these technologies, which will increase the time and resource commitment.
- **Change Management:** Changes in procedures and culture are frequently necessary when introducing new technologies, which may demand additional expenditure in change management initiatives.

Addressing the challenge

- **Comprehensive ROI Analysis:** Analyze Industry 4.0 implementation's potential return on investment (ROI) in great detail. Quantifying the anticipated advantages, such as greater quality, decreased downtime, and increased productivity, is necessary.
- **Pilot Projects:** Before committing to full-scale adoption, start with small-scale pilot projects to verify the impact of Industry 4.0 technology. Successful pilots might attract additional funding.
- **Phased Approach:** Implement Industry 4.0 technology in stages so that the company can spread costs over time and gain knowledge from the results of each phase before moving on to the next.
- **Strategic Partnerships:** Explore cost-sharing possibilities and expertise by working with technology providers, consultants, and solution integrators.
- **Funding Sources:** Investigate alternative funding options for the adoption of cutting-edge technologies, such as grants, subsidies, tax breaks, and venture capital.
- **Shared Infrastructure:** Work with other businesses or industry groups to split the expense of developing and upgrading your infrastructure.
- **Leasing and Financing:** To stretch out the initial costs over time, think about leasing equipment or looking into financing possibilities.
- **Long-Term Perspective:** Think of your investment in Industry 4.0 as a long-term plan that could eventually result in competitive advantages and higher profits.
- **Risk Management:** Develop measures to reduce any potential risks and uncertainties related to the adoption of Industry 4.0.
- **Benchmarking:** To learn more about the market environment and investment patterns, compare your company's investment in Industry 4.0 with industry benchmarks.
- **Employee Involvement:** Engage staff in decision-making to ensure that investments address pain points and correspond with strategic goals.

Although the initial investment in Industry 4.0 can be significant, businesses that approach the implementation strategically and take the long-term advantages into account are more likely to see a favorable ROI.

Organizations can minimize the financial risks involved with adopting cutting-edge technologies by carefully planning, piloting, and investigating other funding and collaboration alternatives.

CHANGE MANAGEMENT AND CULTURAL SHIFT

Critical difficulties in the context of Industry 4.0 include change management and cultural transition. The term "Industry 4.0" refers to the Fourth Industrial Revolution, which has advanced manufacturing, automation, data analytics, and digital technology at its core. While these developments have many advantages, they also force firms to make considerable adjustments to their procedures, organizational structures, and mentalities. The role of cultural change and change management in Industry 4.0 is as follows:

- **Technological Disruption:** Disruptive technologies like the IoT, artificial intelligence (AI), machine learning, and advanced robots are all part of Industry 4.0. The way operations are run must significantly change to accommodate these technologies, and personnel must pick up new skills and get used to new workflows.
- **Change in Work Processes:** Employees must become used to new ways of working as laborious processes are automated and optimized. Redefining employment roles, responsibilities, and tasks may be necessary to achieve this. Employees may resist change if they believe their jobs are at danger or if they are not sufficiently prepared for the transition.
- **Skill Gap:** A workforce with more tech knowledge is required for Industry 4.0. For its staff to be able to use and maintain the latest technology, organizations must invest in upskilling and reskilling them. To close the skills gap, one must be dedicated to lifelong learning and development.
- **Cultural Shift:** The cultural change required for Industry 4.0 entails promoting an atmosphere of creativity, adaptability, and collaboration. It may be necessary to replace rigid hierarchical structures with more adaptable and agile ones that promote experimentation and cross-functional collaboration.
- **Leadership Support:** The ability to lead is essential for fostering change. Leaders must convey the justification for implementing Industry 4.0, allay worries, and establish the transformation's vision. Their cooperation and participation indicate the seriousness of the changes and foster employee trust.
- **Employee Engagement:** Employee involvement in the change process can lower resistance and boost ownership. An easier transition may result from asking for their opinions, addressing their worries, and including them in the decision-making process.

- **Communication:** During times of change, communication must be clear and effective. Employees need to be aware of the changes that are taking place, what is expected of them, and how their responsibilities may alter in the future. Transparent communication can reduce fear and apprehension.
- **Risk Management:** The introduction of new technology may be fraught with technical difficulties and disruptions. To reduce risks, organizations must have plans in place both before and after the technologies are fully integrated into daily operations.
- **Ethical and Privacy Considerations:** Technologies used in Industry 4.0 sometimes involve handling enormous volumes of data. Concerns regarding data privacy, security, and moral use of technology must be addressed by organizations if they are to foster confidence among stakeholders, including employees and customers.
- **Long-Term Vision:** The transition to Industry 4.0 is ongoing. Organizations must continuously adapt to changing market demands and technological requirements. The ability to see how Industry 4.0 will affect the company's future helps businesses stay focused and committed to change.

In conclusion, tackling challenges [2, 22] with change management and cultural shift is essential for the effective implementation of Industry 4.0 efforts. To meet the difficulties and exploit the opportunities posed by this new industrial revolution, organizations must invest not only in technology but also in their people, processes, and communication strategies.

Example: implementation of AI-powered manufacturing system

As part of its Industry 4.0 transformation, imagine a typical manufacturing company deciding to install an AI-powered manufacturing system. This system makes use of AI algorithms to streamline production procedures, anticipate maintenance requirements, and boost overall effectiveness.

Change management challenges

- **Resistance to Automation:** Automation may be met with resistance from workers who have spent years on the assembly line. They worry about their job security and believe that the new system endangers their jobs.
- **Lack of Skills:** It's possible that the current workforce lacks the expertise needed to run and maintain the AI-powered system. A different skill set, including data analysis, programming, and system troubleshooting, is needed to learn how to deal with AI.

- **Fear of Job Loss:** Employees could worry that the new technology would eliminate their jobs. This anxiety could result in resistance and a lack of desire to accept the change.

Cultural shift challenges

- **Hierarchical Culture:** In the organization's hierarchical culture, senior management traditionally makes choices and communicates them down the chain of command. This culture might prevent open dialogue and teamwork needed for the successful deployment of AI.
- **Lack of Innovation Mindset:** It's possible that the company's culture discourages exploration and innovation. It's possible that staff members are used to following standard operating procedures and are unwilling to explore new technology because of the uncertainty that comes with it.
- **Silos and Lack of Collaboration:** The organization's many departments may operate in isolation from one another and avoid collaborating to implement the new system. This lack of cooperation can prevent the organization from successfully integrating AI.

Addressing the challenges

The organization could implement the following measures to solve these issues with change management and cultural shifts:

- **Clear Communication:** The leadership team must explain the rationale for implementing AI, emphasizing that technology is intended to increase overall efficiency and competitiveness rather than eliminate jobs.
- **Upskilling and Training:** The business should fund training initiatives to assist staff in developing the abilities required to operate the AI-powered system. This expenditure reflects a dedication to supporting employees' professional growth.
- **Inclusive Decision-Making:** Employees at all levels can feel appreciated and as a part of the transition by being involved in decision-making and asking for their views.
- **Innovation Culture:** The business should encourage employees to develop fresh concepts, try out new technology, and work with colleagues from different areas.
- **Change Champions:** Spreading enthusiasm and favorable views about the use of AI can be facilitated by identifying and supporting change advocates inside the organization.

- **Cross-Functional Teams:** To promote collaboration and guarantee that several departments operate harmoniously, cross-functional teams might be established to oversee the implementation.
- **Long-Term Vision:** Employees can gain a broader perspective by being given a long-term vision for how the AI-powered system can boost the business' competitiveness and sustainability.

The business may create an environment where employees are more accepting of the changes brought about by Industry 4.0 and actively participate in its effective implementation by addressing these issues with change management and cultural shift.

DATA OVERLOAD AND ANALYSIS

Significant obstacles that enterprises confront in the context of Industry 4.0 include data overload and analysis problems. Organizations must effectively manage, process, and extract valuable insights from the huge amounts of data produced by Industry 4.0 technologies from a variety of sources in order to make educated decisions. Here is a closer look at the issues with data overload and analysis in Industry 4.0, as well as potential solutions.

Challenges

- **Volume of Data:** IoT sensors, smart devices, and automation, among other Industry 4.0 technologies, produce enormous amounts of data in real time. It might be daunting to manage and process this amount of data.
- **Variety of Data:** In Industry 4.0, data is frequently heterogeneous, originating from many sources and in different formats. Complexity might arise while integrating and interpreting data from several sources.
- **Velocity of Data:** Because of the speed at which data is generated, businesses must process and analyze it immediately in order to gain actionable insights. Traditional analysis techniques could find it difficult to keep up with the speed of data.
- **Veracity of Data:** It can be difficult to guarantee data quality and accuracy. Poor data quality can result in incorrect insights and judgments.
- **Lack of Expertise:** There may be a shortage of data scientists and analysts available to handle and analyze the data in the organizations.
- **Infrastructure:** Strong IT infrastructure, including storage, processing power, and analytics tools, is needed to handle enormous amounts of data.

Solutions

- **Data Strategy:** Organizations should create a thorough data strategy that details the procedures for gathering, storing, processing, and analyzing data. The objectives and resources of the company should be in line with this plan.
- **Data Preprocessing:** Prior to analysis, data must be cleaned, transformed, and organized to ensure data quality. Techniques for preparing data assist in getting rid of contradictions and mistakes.
- **Advanced Analytics Tools:** Automate data analysis processes and extract valuable patterns and insights by using advanced analytics technologies like machine learning algorithms and AI models.
- **Real-Time Analytics:** Utilize real-time analytics solutions to process and examine data as it is being generated, enabling prompt response to changing circumstances.
- **Data Integration:** Utilize data integration systems to bring together information from many sources and produce a uniform perspective, making it simpler to analyze and come to relevant conclusions.
- **Data Visualization:** Data-driven choices are made easier for stakeholders to grasp when data is presented visually through charts, graphs, and dashboards.
- **Predictive and Prescriptive Analytics:** Predictive and prescriptive analytics can be used to examine previous data, predict future patterns, and suggest the best course of action.
- **Cloud Solutions:** Organizations may handle massive volumes of data using cloud computing's scalable processing and storage capabilities without making big infrastructure investments.
- **Skill Development:** Spend money on staff upskilling and training so they can master data analysis and draw insightful conclusions from the data.
- **Data Governance:** To guarantee data quality, security, and compliance, implement data governance methods into practice. It is crucial to have clear rules about data ownership, access, and usage.
- **Partnerships:** Work together with other partners, such as consultants or data analytics companies, to benefit from their knowledge and resources for efficient data analysis.
- **Continuous Improvement:** Review and improve data analysis procedures on a regular basis to accommodate changing data trends and business requirements.

In conclusion, Industry 4.0's problems with data overload and analysis need a strategic approach that includes technology solutions, knowledgeable staff, and a clearly defined data management plan. Organizations may make use of Industry 4.0 technology to spur innovation and boost operational efficiency by properly managing data and collecting meaningful insights.

Example: smart factory monitoring system

Imagine a manufacturing organization that has adopted the ideas of Industry 4.0 and put in place a smart factory management system. To gather real-time data on numerous factors like temperature, pressure, vibration, and production rates, this system leverages IoT sensors installed on machines. The objectives are to increase overall efficiency, decrease downtime, and optimize production processes.

Data overload and analysis challenges

- **High Data Volume:** Each machine in the plant produces a significant amount of data for the IoT sensors. The amount of data generated by hundreds of sensors functioning concurrently becomes overwhelming.
- **Real-Time Data:** Real-time data streams from the sensors create an ongoing information deluge. Real-time processing and analysis of this data might be resource-intensive.
- **Data Variety:** Different sensor types produce various data forms and formats. It can be difficult to integrate and analyze this varied data.
- **Data Velocity:** The real-time nature of the monitoring system causes the data to come in at tremendous speeds. It's possible that traditional analytic techniques can't keep up with how quickly data is generated.
- **Data Quality:** It might be difficult to ensure data quality and dependability because sensor data may contain mistakes, outliers, or inconsistent results.

Potential solutions

- **Data Aggregation:** Aggregate data collected over predetermined time intervals should be examined rather than the raw data from individual sensors. This decreases the volume of data overall while maintaining important insights.
- **Sampling:** To choose a selection of data points for analysis, use sampling procedures.
- Statistical techniques can help guarantee that the sample fairly depicts the overall dataset.
- **Edge Computing:** Reduce the amount of data sent to central computers by processing and analyzing data closer to the data source. This could enhance the efficiency of real-time analysis.
- **Data Filtering:** Apply data filtering techniques to the sensor data before analysis to eliminate noise and outliers.
- **Machine Learning:** To automatically detect patterns and anomalies in sensor data, use machine learning techniques [4]. This can aid in problem detection and maintenance forecasting.

- **Data Prioritization:** Sort data streams according to their importance and potential influence on output. To manage the data overload, concentrate on the important data points.
- **Cloud Analytics:** Utilize cloud-based analytics solutions that offer scalable processing capability to manage high data volumes and conduct intricate analysis.
- **Predictive Maintenance:** Reduce downtime by implementing predictive maintenance algorithms that use historical sensor data to foretell when machines are likely to need maintenance.
- **Data Visualization:** Make dashboards that provide critical performance indicators and alarms based on sensor data in real time. Operators and management benefit from visualization tools' quick data interpretation.
- **Automated Alerts:** Create automatic alarm systems that send messages to operators or maintenance crews when specific thresholds or anomalies are found in the sensor data.
- **Data Governance:** Establish data governance procedures to guarantee the consistency, accuracy, and standardization of your data. Data quality checks must be performed often.

The manufacturing company may efficiently use the insights from its smart factory monitoring system to boost production processes, maximize resource utilization, and make informed decisions in real time by tackling data overload and analytical difficulties.

REGULATORY AND LEGAL CHALLENGES

The rapid evolution of technology creates new complications that call for suitable legal frameworks and laws to ensure safety, privacy, and ethical use, making regulatory and legal difficulties significant in the context of Industry 4.0 [20]. Here are some of the main governmental and legal issues raised by Industry 4.0.

- **Data Privacy and Security:** Data privacy and security are raised by the increase of connected devices and data collection. It becomes difficult to strike a balance between using data for innovation and protecting people's personal information. Protecting people's right to privacy is a goal of laws like the General Data Protection Regulation (GDPR) in the European Union, which mandates that corporations handle data responsibly.
- **Cybersecurity:** Industry 4.0 systems' greater interconnectedness raises the danger of cyberattacks and data breaches [23, 24, 25]. Regulations must address the cybersecurity measures that enterprises must employ to safeguard sensitive data and critical infrastructure.

- **Intellectual Property (IP):** Intellectual property conflicts may arise from the creation and use of novel technologies. In the context of collaborative work and common platforms, determining ownership of algorithms, designs, and solutions can be challenging.
- **Liability and Accountability:** As automation and AI become ingrained in business processes, concerns regarding responsibility for accidents or mistakes brought on by autonomous systems surface. When technology malfunctions or results in unforeseen effects, legal frameworks must establish who is responsible.
- **Ethical Considerations:** Industry 4.0 brings moral conundrums such as AI bias, job displacement brought on by automation, and the use of AI in decision-making. Regulations must address the moral application of technology and guarantee the impartiality, fairness, and openness of AI systems.
- **Cross-Border Data Flow:** Industry 4.0 frequently entails cross-border data exchange. Different data protection laws in different countries can make data storage and movement more difficult.
- **Standardization:** Industry 4.0's absence of established protocols and interfaces may make it more difficult to interoperate and be compatible. To ensure smooth technology integration, regulatory agencies may need to encourage or impose specific standards.
- **Workforce Impact:** Industry 4.0-related automation and digitization may result in job displacement and changes to the nature of the workplace. In this changing environment, regulatory measures may be required to handle retraining, reskilling, and worker rights.
- **Competition and Monopoly Concerns:** Concerns about competition and market dominance might arise from the concentration of power within a small number of technology titans. In the Industry 4.0 ecosystem, anticompetitive behavior may need to be monitored and controlled by regulatory organizations.
- **Environmental Impact:** While improving productivity, Industry 4.0 technology may also have an impact on the environment. In light of these technologies, regulations may need to address resource management and sustainability.
- **Regulation Adaptation:** Regulatory adaptation frequently lags behind the quick speed of technological advancement. To effectively address new challenges and develop technology, legal frameworks must be flexible.

Governments, industry stakeholders, legal experts, and technology developers must work together to address these regulatory and legal difficulties. The goal is to promote innovation while making sure that Industry 4.0 technologies are used in a responsible, ethical, and secure manner for the good of society as a whole.

MAINTENANCE AND UPGRADES

The integration of cutting-edge technologies, automation, and networked systems necessitates careful planning, execution, and continuous management to maintain operational efficiency, dependability, and durability, making maintenance and upgrades crucial concerns in the context of Industry 4.0. Following are some of the primary issues to be aware of when maintaining and upgrading in Industry 4.0.

Challenges

- **Complexity of Systems:** Industry 4.0 systems frequently consist of a complicated network of interconnected hardware, software, sensors, and machines. Such complex systems demand specialist knowledge for maintenance and upgrades.
- **Diverse Technologies:** IoT devices, robotics, AI, and cloud computing are just a few of the many technologies that are incorporated within Industry 4.0. Each technology has particular upkeep and upgrade needs.
- **Predictive Maintenance:** Industry 4.0 makes predictive maintenance possible through data analysis, but putting predictive maintenance techniques into practice effectively necessitates reliable data collection, analysis, and prompt actions.
- **Downtime Minimization:** Operations might be disrupted and severe financial losses can result from downtime in Industry 4.0 systems. It becomes difficult to coordinate maintenance and upgrades with production schedules.
- **Interoperability:** The compatibility of other system components may be impacted by upgrading one component. To prevent interruptions, it is crucial to ensure seamless integration throughout upgrades.
- **Skill Gap:** Updating and maintaining Systems based on Industry 4.0 require a workforce with expertise in a variety of fields, such as data analytics, cybersecurity, robotics, and AI. Finding and keeping qualified workers can be difficult.
- **Legacy Systems Integration:** It can be challenging to integrate Industry 4.0 technology with current legacy systems, necessitating careful planning and perhaps phased upgrades.

Considerations and solutions

- **Life Cycle Management:** Implement a thorough lifetime management strategy that includes periodic upgrades, routine maintenance, and retirement plans for outdated systems.
- **Remote Monitoring:** Connectivity and IoT sensors can be used to enable remote equipment health monitoring. This eliminates the requirement for manual inspections and enables preventive maintenance.

- **Predictive Analytics:** Utilize data analytics to foretell the need for maintenance based on machine performance and health data. This lessens unexpected downtime.
- **Condition Monitoring:** Utilize real-time condition monitoring to keep track of elements such as pressure, vibration, and temperature. Any irregularities could send maintenance teams notice.
- **Modularity:** Design systems with modularity in mind to make component upgrades and replacements easily without harming the system as a whole.
- **Scalable Infrastructure:** Create infrastructure that can expand with new technology. Upgrades can be made with scalability and flexibility using cloud-based technologies.
- **Standardization:** Adopt industry standards for parts and protocols to make system updates easier and to ensure compatibility.
- **Continuous Training:** To keep maintenance crews up-to-date on the newest technology and best practices, invest in ongoing training.
- **Cybersecurity:** You should make sure that upgrades don't jeopardize cybersecurity. Update security procedures frequently, fix flaws, and safeguard against online dangers.
- **Vendor Relationships:** Keep solid connections with technology vendors. They can offer assistance, updates, and perceptions into recommendations for upkeep and improvements.
- **Pilot Projects:** Before deploying new technology widely, think about starting with pilot projects to test upkeep and upgrade methods.
- **Feedback Loop:** Create a feedback loop where maintenance workers may inform developers about problems encountered during upkeep and updates. This could result in system design advancements.

In summary, Industry 4.0 maintenance and upgrades demand a deliberate strategy that strikes a balance between operational requirements, technical developments, and risk reduction. In today's changing industrial environment, effective maintenance and timely upgrades are crucial for guaranteeing the dependability, performance, and life span of Industry 4.0 equipment.

ENERGY CONSUMPTION AND SUSTAINABILITY

Sustainability and energy use are important factors to take into account when implementing Industry 4.0 technologies. While Industry 4.0 has many advantages in terms of production and efficiency, it also poses problems with regard to energy use and environmental effects. Here is a closer look at the problems with and remedies for Industry 4.0's energy use and sustainability.

Challenges

- **Increased Data Processing:** Industry 4.0 technologies produce and process enormous amounts of data, thus requiring a lot of computer resources. Infrastructure for data centers and cloud services may result in increased energy use.
- **Automation and Robotics:** Although automation can increase efficiency, it is possible that it will use more energy than manual labor because it involves machines and robots.
- **IoT Devices:** Due to the constant use of sensors, connectivity modules, and communication networks, the proliferation of IoT devices in Industry 4.0 may result in higher energy consumption.
- **Waste Generation:** The accelerated pace of technology development may cause equipment and devices to quickly become obsolete, which would increase electronic waste and deplete resources.
- **Resource-Intensive Materials:** Advanced technologies like semiconductors and high-performance materials demand a lot of energy and resources, which has an impact on the environment.

Considerations and solutions

- **Energy-Efficient Designs:** Design devices, sensors, and machines for Industry 4.0 systems that are energy-efficient. Optimize algorithms and use low-power components to cut back on energy use.
- **Renewable Energy Sources:** Utilize renewable energy sources like solar, wind, and hydropower to power Industry 4.0 processes. Changing to clean energy helps reduce carbon emissions and has a positive influence on the environment.
- **Smart Energy Management:** Implement smart energy management systems that continuously monitor energy use and adjust consumption patterns in response to demand and supply.
- **Energy Harvesting:** Investigate energy-harvesting methods to use sources like ambient light, vibration, or heat to power low-energy IoT devices. This lessens the need for outside power sources.
- **Efficient Data Centers:** Utilize server virtualization, energy-saving technology, and efficient cooling systems to optimize data center operations. Data centers are essential for lowering energy use.
- **Demand Response:** Utilize demand response tools to modify energy use in accordance with times of high demand and grid circumstances, helping to maintain grid stability.
- **Life Cycle Assessment:** Conduct life cycle analyses to determine how Industry 4.0 technologies affect the environment from production to disposal. Utilize these evaluations to guide your judgments about sustainable design.

- **Circular Economy:** Adopt circular economy concepts by creating items that are durable, repairable, and recyclable. Reduce waste and prolong the life of gadgets.
- **Collaboration and Standards:** Establish energy efficiency standards and recommendations for Industry 4.0 technologies in cooperation with partners in the industry, with regulators, with governments, and with other organizations.
- **Employee Engagement:** Encourage energy-efficient behaviors inside the company and raise employee knowledge of energy-saving programs to get them involved.
- **Innovation for Sustainability:** Encourage the development of new sustainable technologies that can be included in Industry 4.0 systems, such as sensors, materials, and procedures that are energy-efficient.
- **Regulatory Compliance:** Keep current with environmental standards and laws that affect sustainability and energy use. Make sure that all applicable laws and standards are followed.

To minimize the environmental impact of technology breakthroughs, Industry 4.0 strategies must take sustainability and energy-efficient practices into account. Organizations may help create a more sustainable future by emphasizing energy efficiency and implementing ethical standards while utilizing Industry 4.0 technologies.

RURAL AND SMALL-SCALE ADOPTION

Due to issues including constrained resources, infrastructure, and awareness, the adoption of Industry 4.0 technologies in rural and small-scale settings creates particular difficulties. Although the deployment of Industry 4.0 has the potential to significantly help these areas, there are a number of difficulties that must be resolved.

Challenges

- **Infrastructure Limitations:** The availability of dependable internet connectivity, which is essential for data transmission and communication among linked devices and systems, is sometimes limited in rural locations.
- **Cost Constraints:** Small enterprises and rural areas might not have the money to invest in the digital infrastructure and tools needed for Industry 4.0 adoption.
- **Lack of Awareness:** Adoption may be hampered by a lack of knowledge about and understanding of Industry 4.0 technologies. Change

may be met with resistance because people are unsure of its advantages and potential effects.

- **Skill Gap:** The trained labor required to install, run, and maintain Industry 4.0 technologies, such as data analysis and software development, may not be present in rural and small-scale settings.
- **Customization Challenges:** Smaller businesses could have particular processes that call for specialized solutions. It can be difficult to implement Industry 4.0 solutions that are already available.
- **Regulatory Barriers:** It's possible that the regulatory environments in rural and small town settings will make it difficult to employ cutting-edge technology. Regulation adherence can be difficult.

Considerations and solutions

- **Awareness and Education:** Educate the public about the advantages of Industry 4.0 through workshops, training courses, and public awareness initiatives. By informing interested parties of the benefits, we can encourage interest and adoption.
- **Adaptation to Local Needs:** Create Industry 4.0 solutions that are specifically suited to the requirements and working methods of rural and small-scale enterprises. Customized solutions can increase the likelihood of acceptance.
- **Collaborative Initiatives:** To pool resources and knowledge for successful adoption, and encourage cooperation between technology suppliers, governmental organizations, educational institutions, and local communities.
- **Public-Private Partnerships:** To share the costs and risks of deploying Industry 4.0 technologies in rural areas, and create collaborations between the public and commercial sectors.
- **Incentives and Subsidies:** To promote adoption among rural and small businesses, governments and organizations might offer incentives, grants, and subsidies. These financial aids might lessen the strain of the initial investment.
- **Skill Development:** Provide workshops and training programs to equip the local workforce with the skills needed to effectively use and maintain Industry 4.0 technologies.
- **Technology Adaptation:** Start with Industry 4.0 solutions that are easier to use and more accessible before adding more complex ones gradually. This method ensures gaining familiarity and assurance gradually.
- **Shared Facilities:** Create collaborative spaces or technology hubs that various small enterprises can use and access.
- **Digital Inclusion:** Improve digital infrastructure and internet access in rural areas to facilitate the industry's adoption and integration.

- **Pilot Projects:** Start with small-scale pilot projects to illustrate the advantages of Industry 4.0. Positive results from these initiatives could promote wider adoption.
- **Long-Term Vision:** Create a long-term blueprint for rural adoption of Industry 4.0. This road map should outline phases for integrating technology, developing skills, and enhancing infrastructure.

Rural and small-scale firms can overcome obstacles to Industry 4.0 adoption and use technology to increase productivity, efficiency, and competitiveness in their particular contexts by addressing these issues and putting forth customized solutions.

REFERENCES

1. Li, X., Zhang, W., Zhao, X., Pu, W., Chen, P., & Liu, F. (2021). Wartime industrial logistics information integration: Framework and application in optimizing deployment and formation of military logistics platforms. *Journal of Industrial Information Integration*, 22, 100201.
2. Llovet, R., Eisenhauer, J. J., & Schmiech, E. L. (2022). New build deployment challenges and resolution. In *Fundamental Issues Critical to the Success of Nuclear Projects* (pp. 195–220), by Joseph Boucau Woodhead Publishing.
3. Jiagui, X. I. E., Chao, Q. I., & Jiajia, Z. H. U. (2020). Architecture and deployment progress of identifier resolution system for Industrial Internet. *Information and Communications Technology and Policy*, 46(10), 10.
4. Qiu, H., Vavelidou, I., Li, J., Pergament, E., Warden, P., Chinchali, S., ... Katti, S. (2022). ML-EXray: Visibility into ML deployment on the edge. *Proceedings of Machine Learning and Systems*, 4, 337–351.
5. Majid, M., Habib, S., Javed, A. R., Rizwan, M., Srivastava, G., Gadekallu, T. R., & Lin, J. C. W. (2022). Applications of wireless sensor networks and internet of things frameworks in the industry revolution 4.0: A systematic literature review. *Sensors*, 22(6), 2087.
6. Sherwani, F., Asad, M. M., & Ibrahim, B. S. K. K. (2020, March). Collaborative robots and industrial revolution 4.0 (IR 4.0). In 2020 International Conference on Emerging Trends in Smart Technologies (ICETST) (pp. 1–5). IEEE.
7. Alazab, M., Gadekallu, T. R., & Su, C. (2022). Guest editorial: Security and privacy issues in industry 4.0 applications. *IEEE Transactions on Industrial Informatics*, 18(9), 6326–6329.
8. Ustundag, A., Cevikcan, E., Ervural, B. C., & Ervural, B. (2018). Overview of cyber security in the industry 4.0 era. In B. C. Ervural & B. Ervural (eds.), *Industry 4.0: Managing the Digital Transformation* (pp. 267–284). Springer.
9. Qahtan, S., Sharif, K. Y., Zaidan, A. A., Alsattar, H. A., Albahri, O. S., Zaidan, B. B., ... & Mohammed, R. T. (2022). Novel multi security and privacy benchmarking framework for blockchain-based IoT healthcare industry 4.0 systems. *IEEE Transactions on Industrial Informatics*, 18(9), 6415–6423.
10. Hozdić, E. (2015). Smart factory for industry 4.0: A review. *International Journal of Modern Manufacturing Technologies*, 7(1), 28–35.

11. Harrison, R., Vera, D., & Ahmad, B. (2016). Engineering the smart factory. *Chinese Journal of Mechanical Engineering*, 29(6), 1046–1051.
12. Lee, G. Y., Kim, M., Quan, Y. J., Kim, M. S., Kim, T. J. Y., Yoon, H. S., ... Ahn, S. H. (2018). Machine health management in smart factory: A review. *Journal of Mechanical Science and Technology*, 32, 987–1009.
13. Davis, J., Edgar, T., Graybill, R., Korambath, P., Schott, B., Swink, D., ... & Wetzel, J. (2015). Smart manufacturing. *Annual Review of Chemical and Biomolecular Engineering*, 6, 141–160.
14. O'Donovan, P., Leahy, K., Bruton, K., & O'Sullivan, D. T. (2015). An industrial big data pipeline for data-driven analytics maintenance applications in large-scale smart manufacturing facilities. *Journal of Big Data*, 2(1), 1–26.
15. Chen, T., & Lin, Y. C. (2017). Feasibility evaluation and optimization of a smart manufacturing system based on 3D printing: A review. *International Journal of Intelligent Systems*, 32(4), 394–413.
16. Tao, F., Qi, Q., Liu, A., & Kusiak, A. (2018). Data-driven smart manufacturing. *Journal of Manufacturing Systems*, 48, 157–169.
17. Chen, B., Wan, J., Shu, L., Li, P., Mukherjee, M., & Yin, B. (2017). Smart factory of industry 4.0: Key technologies, application case, and challenges. *IEEE Access*, 6, 6505–6519.
18. Mabkhot, M. M., Al-Ahmari, A. M., Salah, B., & Alkhalefah, H. (2018). Requirements of the smart factory system: A survey and perspective. *Machines*, 6(2), 23.
19. Wang, S., Wan, J., Li, D., & Zhang, C. (2016). Implementing smart factory of industrie 4.0: An outlook. *International Journal of Distributed Sensor Networks*, 12(1), 3159805.
20. Habrat, D. (2020). Legal challenges of digitalization and automation in the context of industry 4.0. *Procedia Manufacturing*, 51, 938–942.
21. Agrawal, S., Sahu, A., & Kumar, G. (2022). A conceptual framework for the implementation of industry 4.0 in legal informatics. *Sustainable Computing: Informatics and Systems*, 33, 100650.
22. Ada, N., Kazancoglu, Y., Sezer, M. D., Ede-Senturk, C., Ozer, I., & Ram, M. (2021). Analyzing barriers of circular food supply chains and proposing industry 4.0 solutions. *Sustainability*, 13(12), 6812.
23. Fernandez-Carames, T. M., & Fraga-Lamas, P. (2019). A review on the application of blockchain to the next generation of cybersecure industry 4.0 smart factories. *IEEE Access*, 7, 45201–45218.
24. Singh, R., Akram, S. V., Gehlot, A., Buddhi, D., Priyadarshi, N., & Twala, B. (2022). Energy system 4.0: Digitalization of the energy sector with inclination towards sustainability. *Sensors*, 22(17), 6619.
25. Javaid, M., Haleem, A., Singh, R. P., Khan, S., & Suman, R. (2022). Sustainability 4.0 and its applications in the field of manufacturing. *Internet of Things and Cyber-Physical Systems*, 2, 82–90.
26. Gumz, J., Fettermann, D. C., Frazzon, E. M., & Kück, M. (2022). Using industry 4.0's big data and IoT to perform feature-based and past data-based energy consumption predictions. *Sustainability*, 14(20), 13642.

Chapter 11

Hate speech detection using machine learning models

Pradeep Gupta and Sonam Gupta

INTRODUCTION

Academics and practitioners in the field of identifying and flagging discriminatory can benefit from this study as it presents a comprehensive and detailed review of the latest advancements in this area. With the growing use of social media, the incidence of inappropriate language has grown, enabling it to spread quickly and reach a large audience, causing significant harm to those targeted.

The impact of offensive language on social media platforms can be far-reaching and serious, including psychological harm to the targets, degradation of civil discourse, and the promotion of discriminatory attitudes and behaviors. To solve this issue, researchers and technology firms are working on ML algorithms to detect discrimination on social media platforms, with the objective of automating the process of finding and eliminating damaging information. However, identifying hate speech is a difficult undertaking since it frequently entails context-specific language use, cultural variations, and subjective interpretations.

In recent times, hate speech has emerged as a widespread issue, owing to the widespread adoption of social media and other online communication channels in modern times. With the growth of these platforms, hate speech has also spread, leading to serious consequences such as division, discrimination, and violence. The harm caused by hate speech is not limited to individuals and groups who are targeted but also extends to society as a whole by affecting social cohesion and trust.

To address the increasing concern surrounding hate speech, scholars and innovators have been collaborating to design automated systems for detecting hate speech that possess the capability to precisely recognize and categorize hate speech in online texts. These automated solutions have the potential to play an important role in combating hate speech and building a safer and comprehensive online environment.

NLP activities, including text categorization, sentiment analysis, and speech recognition, typically employ ML and DL algorithms. These approaches can be employed to tackle the issue of identifying hate speech, as they have

DOI: 10.1201/9781003479031-11

been demonstrated to be potent in comprehending patterns and associations between words and phrases in a text and can be trained on sizable datasets to heighten precision.

Moreover, ML algorithms must be carefully designed to avoid over-blocking or under-blocking of content and to balance the protection of free speech with the need to curb hate speech. Despite the difficulties, the use of ML to track and analyze online conversations to identify hate speech continues to be a lively research area, with the aim of devising more efficient and potent resolutions to alleviate the harm caused by this form of communication.

NLP, DL, and rule-based algorithms are among the techniques utilized in ML-based hate speech identification. The NLP techniques employed involve named entity identification to classify text as expressing hateful sentiments or not, text classification, and sentiment analysis. DL models, such as CNNs and RNNs, are implemented to comprehend the representations of text and recognize hate speech in a more sophisticated manner. In contrast, rule-based methods rely on a predefined set of rules to detect hate speech but are less adaptable and necessitate manual updates to keep up with emerging forms of hate speech.

Hate speech detection systems that are based on ML techniques are typically evaluated using common metrics such as accuracy, precision, and F1 score. These measurements show the ratio of false positives to false negatives. However, depending on the standard and quantity of information used for training, the level of detail of the algorithm, and the domain-specific aspects of the language, the performance of these systems may vary.

In general, the application of ML to detect hate speech on social networks has demonstrated good outcomes in identifying and preventing hate speech. Nonetheless, there are still obstacles to overcome, such as the skewed nature of the training data, the subjective definition of hate speech, and the ethical and legal implications of automated censorship. To ensure the equitable and successful use of these systems, it is necessary to address these obstacles.

Various types of ML algorithms are utilized in hate speech detection. These include:

1. **Supervised Learning Algorithms:** To categorize text into different classifications, for example, hateful speech or non-hateful speech, the aforementioned algorithms utilize labeled training data. This process is known as supervised learning, and it employs techniques such as DT, RF, and SVM.
2. **Unsupervised Learning Algorithms:** Unsupervised learning algorithms detect patterns and structure in data without the aid of labeled training data. Clustering algorithms and dimensionality reduction techniques are two examples of unsupervised learning algorithms.
3. **Semi-supervised Learning Algorithms:** Semi-supervised learning algorithms are ML techniques that employ data that is both labeled and unlabeled. These algorithms can be used to make predictions or classify text. In hate speech detection, semi-supervised learning algorithms can be especially useful because obtaining a large labeled dataset can be challenging.

This comprehensive analysis serves a dual purpose: first, to assess the current status of hate speech detection, and second, to identify the most efficient algorithms and datasets that have been employed in this domain. The primary reason for the survey is to offer a complete summary of the latest developments and act as a useful reference for practitioners and researchers working within the scope of hate speech identification. By conducting a methodical analysis of the prior research, we hope to obtain insights into the optimal methods for detecting hate speech.

Research Questions

- What are the most commonly used tactics for identifying hate speech on social media, and how do their accuracy and efficiency compare?
- What are the various ways in which hate speech is expressed on social media, including but not limited to racism, sexism, and homophobia, and what obstacles are encountered in detecting these forms of hate speech using automated methods?
- What are the limitations of current hate speech detection algorithms, and how can they be improved to better detect and remove hate speech from social media?

Section 11.1 of the study aims to define hate speech and its societal impact. Section 11.2 reviews related literature in the field, evaluating the diverse techniques and algorithms employed for hate speech observation. Section 11.3 examines the research questions in detail. Lastly, in Section 11.4, the research findings are summarized, and a definitive conclusion is reached, emphasizing the potential of ML & DL algorithms in detecting digital hate speech. The study also highlights the challenges and limitations that require further investigation in future research.

LITERATURE REVIEW

In a specific research project, the efficacy of the SVM algorithm for hate speech recognition was investigated [1]. The study discovered that using bigram characteristics produced the greatest results, with an overall accuracy of 79%. This study is noteworthy because it lays the foundation for future research into autonomous text categorization systems based on NLP and ML. Furthermore, the study sheds light on the most efficient way for detecting automated hate speech messages.

In their research, the author [2] employed the PRISMA (Preferred Reporting Items for Systematic Reviews and Meta-Analyses) approach to perform a comprehensive and thorough examination of the available literature on the recognition of online vitriol. A total of 31,714 publications from 2015 to 2020 were analyzed, and based on the established inclusion criteria, 41 papers were included in the review. The results of the research suggested that the SVM in ML technique was frequently employed in iden-tifying hostile text, although there was a notable growth in the usage of DL algorithms and hybrid DL approaches. The research showed that the use of ML & DL techniques was effective in detecting abusive language on social media. However, to combat hate speech, it was found that a combination of cross-platform models is necessary, integrating multiple technologies and approaches.

In a recent publication [3], an ensemble learning approach was proposed to mitigate unintended biases in the recognition of hate speech with respect to gender. The work trained a collection of predictors utilizing nine compo-nent spaces, and the efficacy of the proposed approach was tested using a publicly accessible dataset. According to the findings, the suggested strategy outperformed current innovative systems for identifying hate speech based on gender.

In a study [4], the author notes the growing popularity of DL-based methods and the increasing availability of hate speech datasets. The author conducted a comprehensive empirical assessment of shallow and DL meth-ods for detecting hate speech, with a focus on analyzing three commonly employed datasets. The study focused on practical performance indicators such as accuracy, computing efficiency, the usage of pre-trained models, and domain generalization. The objective of the research was to provide practical guidelines for the application of hate speech detection, examine the latest technology, and suggest potential areas for future investigation.

According to a recent study [5], utilizing multi-channel convolutional neural network (MC-CNN) models could enhance the efficacy of recog-nition of offensive speech, particularly for languages with scarce training data and limited resources. According to the study, MC-CNN models can extract better features by considering multiple channels end-to-end using a Word2Vec embedding layer. While traditional single-channel CNN models were previously used for the recognition of harmful speech with decent results, the study found that their performance relies on the kind of speech being detected and the amount of training data. The MC-CNN model pro-posed by the researchers was assessed using a separate dataset for Amharic hate speech and exhibited superior performance compared to single-channel CNN models. However, it did not achieve the same level of accuracy as the baseline SVM model. According to the findings of the study, the MC-CNN model may be deemed an appropriate replacement method for the identifi-cation of prejudiced language in circumstances where data is sparse.

In a recent investigation [6], researchers developed a new ML model named BiCHAT, which was designed to detect occurrences of hateful lan-guage on social media platforms. The model employs a multi-layer neu-ral network architecture to classify tweets as either hostile or normal. The architecture's structure contains a BERT layer, an attention-based deep twisted layer, a bidirectional Long Short-Term Memory (LSTM) network with attention, and a softmax layer at the end. The study evaluated the performance of the BiCHAT model using three benchmark datasets from Twitter and found that it outperformed both state-of-the-art and baseline techniques, achieving higher accuracy, recall, and f-score. In addition, the contribution of different neural network components to the model's performance was examined, and the study revealed that removing the deep convolutional layer had the most significant effect.

In a recent study [7], the author addressed the subject of identifying instances of hateful language on networking sites used in Arab countries. As the number of Arab social media users grows, stakeholders are becoming increasingly worried about the prevalence of cyber hate speech. To address this issue, the author suggests a unique method that combines personality learning with hate speech identification. The suggested method is intended to extract personality characteristic elements that are critical for recognizing hate speech. The author conducted trials to evaluate the performance of the suggested technique and acquired a macro-F1 score of 82.3%, which beats earlier work. The study emphasizes the significance of cyber hate speech in Arab social media and recommends a new way of dealing with it. The suggested strategy, which combines personality learning with hate speech detection, is demonstrated to be useful in detecting hate speech, as indicated by the model's enhanced performance. The study's findings have substantial implications for stakeholders in the Arab social media environment, and the incorporation of personality characteristics as a component in hate speech detection algorithms offers the potential for future investigation.

The aim of the research [8] was to conduct a thorough examination of the various definitions of prejudiced language as well as traditional approaches to detecting offensive speech in literature. The author surveyed cutting-edge hate speech detection systems, including handmade feature-based and DL-based algorithms. The paper also gave a full review of common benchmark datasets for detecting hate speech and inflammatory language, as well as the methodologies used to attain high classification results. In addition, the study examined multiple performance metrics for hate speech identification and demonstrated the classification scores of commonly used techniques. This study distinguishes itself from previous surveys by providing a thorough and systematic analysis of the challenges and current evaluation methodologies associated with detecting hate speech across different modes of communication and languages. The study offers valuable insights into the current state of hateful speech detection, including definitions, algorithms, datasets, and evaluation metrics, making it a great resource for field investigators and practitioners alike.

The study presented in this work [9] aims to identify multiple aspects of offensive speech using ensemble learning architectures. The authors offer a method for classifying text into several categories, such as "identification hatred," "threatening," "offends," "inappropriate," "toxicity," and "drastic toxic," by combining a pre-trained BERT model with DL models. To achieve this, the authors use two popular word embedding techniques, GloVe and FastText, and combine them with DL models such as Bi-LSTM and Bi-GRU. Prior to integration with BERT for identifying hate speech on social media, these models undergo separate training using annotated datasets containing instances of offensive language. The ROC-AUC score of 98.63% in the research illustrates the efficacy of the proposed strategy. This demonstrates the approach's potential for enhancing the accuracy of racist language identification on digital communication. Overall, this work sheds light on the utilization of ensemble learning architectures for multi-aspect hate speech identification as well as the possibility of contemporary word embedding approaches for enhancing hate speech detection accuracy.

In this research, the author [10] conducted analysis of discriminatory language in English, compiling a dataset of 451,709 sentences. Out of this dataset, 371,452 sentences were classified as hate speech, while the remaining 80,250 sentences were classified as non-hate speech. To balance the dataset, the author utilized data augmentation techniques to create a balanced dataset with a total of 726,120 samples. In addition, the author developed a custom vocabulary consisting of 145,046 words, which included 6,403 contractions and 377 bad words that are frequently used in hateful content. To ensure the dataset was suitable for training and cross-validation, the maximum allowed word count for each sentence was set to 180 words. The resulting dataset of hate speech and contractions has significant potential for NLP data preparation projects. By expanding the dataset, it is possible to decrease out-of-context phrases. The dataset containing instances of hateful language can also be utilized to build hate speech classifiers for social networking platforms.

The author of this research article [11] introduces a novel algorithm, HaterNet, which can identify and monitor instances of hate speech on Twitter. This expert system was built in partnership with the Spanish National Office against Hate Crimes and expanded by including a new model, HaterBERT, which is based on the BERT architecture. Furthermore, the researchers present a novel methodology referred to as SocialGraph, which builds a user database using a relational network structure, allowing for the analysis of both textual and centrality characteristics. By combining these two approaches, the final model, SocialHaterBERT, surpasses the limitations of simple textual analysis and shows significant improvements in results. This study takes a fresh look at linked models from both a diachronic and dynamic standpoint, pushing beyond the traditional emphasis on textual analysis alone. This opens up new study possibilities and has the potential to improve our understanding of recognition of intolerant speech on internet communities.

In this research [12], the author investigates the impact of harsh language on online platforms and its repercussions on society, with a focus on the function of moralized language in its spread. According to the study, the usage of moralized language is a significant predictor of the incidence of hateful rhetoric on web-based platforms. The author collects three datasets of social media postings and reactions from societal leaders within the disciplines of politics, news media, and advocacy to validate this hypothesis. In this investigation, the researchers utilized both textual analysis and ML techniques to explore the correlation between the presence of the study focuses on examining the presence of honorable and honorable-emotional language in the original tweets and its correlation with the probability of receiving responses containing hateful or abusive language. The study's findings show that a higher frequency of decent and decent-emotional expressions in the original tweets is consistently connected with a higher risk of getting hate speech. More specifically, every new moral phrase raises the likelihood of getting hate speech by 10.76%–16.48%, while every additional moral-emotional term raises the likelihood by 9.35%–20.63%. Overall, the study shows that the use of moralized language is a reliable predictor of derogatory language in internet communities, shedding light on the role of moral and moral-emotional language in the propagation of obscene language online.

The aim of the research performed by the author [13] was to investigate the sentiment analysis of airline data on Twitter by utilizing both text and emoticons. For the analysis, the study used a variety of characteristics. The techniques utilized in the study encompassed TF-IDF, Bag of Words, N-gram, and dictionaries of icons. The findings revealed that the sentiment conveyed by emoticons significantly impacted the sentiment conveyed by text analysis, with emoticons being more dominant in sentiment analysis when utilized. Furthermore, the study revealed that DL algorithms outperformed traditional ML algorithms in sentiment analysis, underscoring the potential of DL in sentiment analysis applications.

In a recent work, the author [14] aimed to tackle some of the challenges linked with the automatic identification of offensive speech on Twitter, including the absence of a standardized framework, issues with precision, challenges with threshold selection, and problems with fragmentation of data issues were among the obstacles. To mitigate these concerns, the author suggested an innovative method for identifying hate speech on Twitter, which involved utilizing a probabilistic clustering model. The proposed approach involved employing a metadata extraction tool to extract tweets that include keywords associated with hate speech, followed by their categorization into either hate speech or non-hate speech using expert categorization based on crowdsourcing. To analyze the tweets further, the author utilized the Term Frequency-Inverse Document Frequency (TF-IDF) model, which was complemented by themes identified using a Bayes classifier. The researchers used a clustering technique that relies on rules to sort real-time tweets into different subject categories, and they utilized fuzzy logic to categorize hate speech. The researchers utilized semantic fuzzy rules in conjunction with a score computation module to detect instances of hate speech. With an F1-score of 0.9256 and an AUC of 0.9645, the proposed model outperformed comparable models. The results were validated using 5-fold cross-validation and a Paired Sample t-Test. Finally, the author put forward a novel probabilistic clustering technique to improve hate speech detection on Twitter, which outperformed previous algorithms.

The author of a recent study [15] managed the difficulties of automated offensive speech classification on Twitter, including the absence of a standardized metadata framework, imprecision, challenges in threshold setting, and fragmentation. The author of a recent study addressed that the task of automating hate speech classification on Twitter is fraught with challenges, which include the lack of a uniform metadata framework, imprecision, cutoff setting problems, and splintering difficulties. To overcome these concerns, in a directive to identify discriminatory remarks on Twitter, the author presented a probabilistic clustering model. This model utilized a metadata extractor to gather tweets containing hate speech keywords, and these tweets were labeled as hate speech or non-hate speech using crowd-sourced expertise. The tweets were then represented using the TF-IDF model, which was augmented with subjects predicted using a Bayes classifier. To categorize real-time tweets into subject groups, a rule-based clustering approach was utilized, and A hate speech categorization system based on fuzzy logic, a score estimate module and linguistic fuzzy rules were created. The suggested model was shown to perform better than comparable models, with an F1-score of 0.9256 and an AUC of 0.9645, as validated by a 5-fold cross-validation and Paired Sample t-Test. The author also provided the proposed generic metadata architecture that provides a framework for hate speech categorization on Twitter by organizing and analyzing metadata such as hashtags, user information, and text features, which had an AUC of 0.97 with accuracy, precision, recall, and F1-score of 0.95, 0.93, 0.92, and 0.93, respectively. The suggested method was found to be effective for automatic topic recognition and classification, and this was statistically verified. Finally, the author presented a full metadata architecture, and the effectiveness of the probabilistic clustering model in categorizing inappropriate words on Twitter was demonstrated, surpassing the performance of similar algorithms.

The study's author [16] presented a multi-task learning framework to enhance individual task performance by exploiting knowledge from related activities. The framework relies on a shared-private structure that separates between shared and task-specific characteristics. To gauge the value of the juggling model, the author tested it on five datasets and found that it showed promising results when it comes to macro-F1 and weighted-F1 scores. These results indicate that the proposed framework successfully leverages information from related tasks to enhance the performance of individual tasks. Finally, the author suggested a multi-task learning framework that uses knowledge from analogous domains to improve individual task performance and proved its usefulness on five datasets.

In this study, the author [17] focused on the issue of misogynistic behavior on the web, and there is a lot of vitriol spoken in English, Italian, and Spanish. The research was carried out as part of the AMI IberEval 2018 and AMI EVALITA 2018 shared tasks to detect sexism on Twitter. The author aimed to investigate the association between gender-based hatred and other types of insulting language and to examine the attainability of identifying cross-linguistic misogyny context. The author proposed a solution for detecting sexism on Twitter that outperformed existing advanced algorithms in benchmark AMI datasets. The study sheds light on the interactions between various forms of abusive language, demonstrating that misogyny is a different form of aggressive language than sexism. The cross-linguistic experiments yielded promising outcomes, with the suggested joint-learning architecture performing well across languages. The author's proposed approach for identifying misogyny on Twitter outperforms existing innovative algorithms and emphasizes the connection between sexism and other forms of harsh language. The multilingual studies demonstrate that recognizing misogynistic language in diverse linguistic context is feasible, with the suggested joint-learning architecture achieving stable performance across languages.

The author of this study [18] explores sentiment analysis, a crucial topic in NLP and data mining, and presents a novel DL model known as the Attention-based Bidirectional CNN-RNN Deep Model (ABCDM) to overcome the limitations of traditional sentiment analysis models. The ABCDM employs bidirectional LSTM and GRU layers to capture prior and forthcoming circumstances, along with a focus mechanism, to highlight significant words during sentiment analysis. Convolution and pooling techniques are also utilized to decrease and to address the challenges of varying object positions. The approach focuses on extracting position-invariant features and reducing feature dimensionality. The ABCDM was evaluated on eight datasets, including five review datasets and three Twitter datasets, and it outperformed other Advance DNN models. The findings of the study indicate the ABCDM model's effectiveness in sentiment analysis and its potential practical applications.

The research presented in [19] focuses on sentiment analysis at the organizational level, a technique used by companies to understand the social sentiment toward their product or service. The use of sentiment analysis extends beyond technical companies; as even non-technical firms can benefit from it by interpreting customer feedback. This is achieved through the application of NLP algorithms, which allow for the digital gathering and analysis of customer reviews. Although the specifics of organizational-level sentiment analysis are mostly proprietary and not publicly available, its use is prevalent among top companies. Sentiment analysis can be implemented by any company in any field through a standard interface, making it a versatile tool for understanding customer sentiment.

The challenge of hate speech has a significant impact on the dynamics of online social communities. Despite efforts by companies to detect and classify hateful content, success has been limited. While state-of-the-art systems have demonstrated high performance in specific datasets, primarily those in English, the generalizability of these methods to other datasets remains uncertain. This study [20] examines the experimental approach employed in previous work and assesses its generalizability to additional datasets. The authors highlight a number of methodological flaws and dataset biases that have resulted in overstated performance claims, mostly as a result of data excessive fitting and random problems. The authors conducted cross-lingual experiments on datasets in both English and Spanish to present a more detailed understanding of current latest methods. This study throws light on the limits of current hate speech detection systems and emphasizes the need for more research in this area.

Various automated algorithms based on NLP, ML, and DL have been presented in recent years for recognizing abusive content on social media networks. A survey [21] gives a thorough examination of the most recent advance techniques for identifying abusive material. The survey classifies the many techniques and characteristics employed by researchers in this subject and indicates the significant obstacles that remain. This report gives useful information about the current situation of the subject and indicates future research topics for building more powerful algorithms to detect abusive content on social media. The authors want to progress this topic by giving a complete examination of present approaches and analyzing the possibilities for future advances in the identification of abusive content.

The study by the authors [22] introduces a "DL-NLP" model designed to detect toxic content on social media platforms. The model is constructed using a blend of convolutional and recurrent layers and is applied on the HASOC2019 corpus. The authors illustrate the model's performance by getting a promising macro F1 score of 0.63. Furthermore, the authors address the issue of overfitting caused by a lack of training data by investigating several ways for increasing the resource pool, such as using unlabeled data and similarly labeled corpora. The results of these experiments show significant improvements in the classification score achieved by the model. This study enhances the science of hate speech identification and indicates the possibility for further advancement.

The author [23] presents a novel technique to sentiment analysis in this work that integrates both text and emoticons. The study examines airline data from Twitter using multiple ML and DL techniques, including Bag of Words, N-gram, TF-IDF, and emoticon lexicons. According to the conclusions of the study, emoticons transmit more sentiment than textual data analysis. Furthermore, the study shows that DL algorithms beat ML systems in text and emoticon sentiment analysis. These findings advance the area of sentiment analysis and have implications for enhancing sentiment analysis model accuracy.

This paper [24] offers a unique approach for detecting hate speech that is built on a deep multi-task learning (MTL) framework. The recommended method aims to tackle the problem of inadequate labeled data by leveraging insights from various connected categorization tasks to enhance the performance of each specific task. The shared and private layers of the multi-task paradigm collect shared and task-specific characteristics, respectively. Experiments on five datasets suggest that the proposed framework is effective in detecting hate speech, as seen by promising macro-F1 and weighted-F1 results. Overall, this work sheds light on the potential of MTL for spotting hateful language and emphasizes its use in overcoming the issue of scarce labeled data.

This article [25] reviews and critiques scholarly studies on Bigotry, offensive language, and social media. The report emphasizes the need for more comprehensive research that goes beyond text-based assessments of overt racist discourse on Twitter in the United States. Rather, the authors advocate for research that investigates broader geographical settings, multiplatform studies, and the visual character of racism on social media. In addition, the study emphasizes the need for reflexivity in research designs in order to prevent promoting colorblind ideas within the sector. The authors also recommend adopting critical race views in order to investigate the underlying workings of social media sites. Finally, the study criticizes the limitations of relying on "hate speech" to approach the moderation and control of racist content and proposes looking into developing work that uses indigenous critical perspectives to investigate racial problems on social media.

This research [26] examines the issue of social networking sites with poisonous content, with a particular emphasis on their prevalence in Arabic literature. Hate speech can cause harm and social tension, but social networks cannot control all user-generated content. As a result, there is a demand for automatic identification of hate speech, particularly in complicated languages such as Arabic. The report examines hate speech in its entirety, including cyberbullying, abusive and insulting language, radicalization, and other anti-social actions. In addition, the paper investigates text mining techniques that can be used for social media analysis. It also discusses the challenges in implementing an Arabic hate speech detection model and offers recommendations for future work. The authors plan to collect data from Twitter and Facebook to train a neural network capable of detecting and classifying Arabic hate speech into distinct categories using the latest DL architectures. In general, the aim of this research is to support the progress of effective techniques for spotting and combating hate speech on social media, with a specific focus on the Arabic language.

The study discussed in [27] is centered on the detection of harassment on social media sites as well as the identification of vulnerable populations. The authors highlight the expanding occurrence of hate speech on social media platforms and its harmful impact on historically marginalized communities. Their suggested approach involves using the Apache Spark parallel computing platform to spontaneously gather and preprocess posts, identify attributes utilizing vocabulary segments and context modeling methods such as Word2Vec, and classify using DL algorithms such as GRU and RNN. The research goes on to explore the need of identifying vulnerable populations, particularly ethnic groups, with limited computational resources, and clusters hate terms using algorithms like Word2Vec to forecast prospective target groups. Because there is no publicly available dataset for Amharic texts, the trials are being undertaken in Ethiopia. The research indicates that the proposed technique correctly identifies the Tigre ethnic group as the most sensitive to hate when compared to Amhara and Oromo. According to the authors, recognizing vulnerable populations is critical for protecting them and developing policies and measures to empower and defend these people. Overall, the article advances the field of hate speech detection and At-risk community recognition by presenting a method that may be used with languages that have limited computing resources.

This research [28] focuses on hate speech and its dissemination on online social media sites, including Gab.com. The authors executed a study to explore the dissemination dynamics of postings produced by hateful and non-hateful Gab users, as well as to analyze these users' account and network features. The study discovered that postings created by hateful individuals spread quicker, further, and to a larger audience than ones created by non-hateful people. Furthermore, hostile users are more closely linked to one another. The study takes a cross-sectional look at how hostile people spread hate material on online social media. The study's findings can assist influence efforts to reduce hate speech and its negative consequences.

The rising incidence of nasty and poisonous content published by some social media users has necessitated the development of an effective automatic toxicity identification model; however, a lack of categorized data and existing biases pose significant hurdles in this sector. To mitigate this problem, the authors [29] propose a novel transfer learning strategy that utilizes the pre-trained language model BERT along with fine-tuning techniques for identifying hostile context in social media content. They benchmark their approach against existing methods by evaluating it on open access Twitter datasets that are labeled for racism, sexism, hate speech, or offensive content. Their model can detect that the data annotation and collection procedures may be influenced by subjective biases, leading to a more precise model.

The author discusses the topic of recognizing websites for social networking with noxious content [30], which has grown more widespread with the advent of sites such as Twitter and Facebook. Despite current NLP research attempts to recognize abusive language, this challenge remains unresolved. The study presents a resilient neural architecture that works well in a variety of languages, including English, Italian, and German. The authors proposed a comprehensive empirical inquiry data to gain a deeper comprehending the responsibilities and impacts of the different components employed in the system. They take into account both the architecture, such as LSTM, GRU, and Bi-LSTM, and the feature selection, such as n-grams, social network-specific features, emotion lexica, emoji, and word embedding. The authors conduct their research using three publicly accessible datasets to identify insulting speech on social media platforms in English, Italian, and German. The study provides insights into the effectiveness of various NLP techniques in detecting hate speech on a larger scale, and the outcomes may be beneficial to researchers and professionals working in this domain.

The authors of this study [31] investigate the increase of racial hatred and prejudice directed toward Asian populations throughout the outbreak of the COVID-19 pandemic on social media platforms. Despite its widespread prevalence, little is known about the development of racial hatred during a pandemic and the role of counter-speech in preventing it. To fill this void, the authors developed COVID-HATE, which is the most extensive compilation of anti-Asian hate and counter-speech to date, encompassing a period of 14 months, incorporating over 206 million tweets and a social network comprising over 127 million nodes. They also constructed a new annotated dataset of 3,355 tweets and utilized it to train a text classifier capable of distinguishing between abusive and counter-speech tweets, with an average macro-F1 score of 0.832. The authors conduct a longitudinal study of tweets using this dataset and users and discover that hateful and counter-speech users are actively connected and engaged with one another rather than living in separate societies. They also found that nodes exposed to hostile information in 2020 were very likely to become hateful. Moreover, counter-speech messages dissuade users from being hostile, suggesting the possibility of reducing hatred on social media platforms through public counter-messaging initiatives. The study's findings offer insights into the social issues created by the COVID-19 pandemic and highlight the potential of public counter-messaging initiatives as a response to hate speech. The freely available dataset and code also make them a valuable resource for future research in this field.

This paper [32] examines the phenomenon of deplatforming by social media companies, particularly for extreme anti-establishment actors who have been labeled as 'dangerous individuals' and removed from popular social media platforms such as Facebook, Instagram, Twitter, and YouTube for organized hate. The paper explores the debates surrounding deplatforming, including concerns over censorship and free speech, and examines the efficacy of deplatforming and how the deplatformed individuals and groups are using alternative social media platforms such as Telegram. The research focuses on the consequences of deplatforming on extreme online person-alities, both established and non-established social media platforms, in addition to the internet in its entirety. Finally, the study explores how the deplatforming policies of social media firms affect critical social media research on extreme speech and its viewers on both mainstream and alter-native platforms.

In a recent study [33], the prevalence of inappropriate language and derog-atory remarks on Arabic social media was examined. The authors introduce the Levantine Hate Speech and Abusive Behavior (L-HSAB) Twitter dataset, which is the first dataset of its kind for the Levantine Arabic dialect. The aim of the dataset is to establish a standard for identifying harmful content on the internet in Arabic. The authors offer an in-depth description of the data collection and annotation procedures, which involved three annota-tors identifying tweets based on a set of predefined rules. They also discuss the reliability of the annotations, which were evaluated using agreement measures such as Cohen's Kappa and Krippendorff's alpha. The L-HSAB dataset contains 5,846 tweets categorized as normal, abusive, or hateful. The authors found high levels of agreement among the annotators, which confirms the reliability of the annotations. However, the authors note that achieving consensus on identifying hate speech is challenging since it involves not only following guidelines but also depends on annotators' prior knowledge and personal beliefs. The authors employ ML-based classifica-tion experiments using NB and SVM classifiers to analyze the dataset. The results from both binary and multi-class classification studies demonstrate that NB is superior to SVM. Finally, the authors underscore the importance of creating publicly accessible datasets for the use of derogatory or offensive language in Arabic dialects that are underrepresented, such as Tunisian and Gulf dialects. The L-HSAB dataset is a valuable resource for researchers and practitioners involved in identifying hate speech and abusive language on Arabic social media and makes a significant contribution to the field of Arabic NLP.

In a recent study [34], the authors draw attention to the rise of a new form of hate speech online as a result of the COVID-19 crisis, which has been prevalent on social media platforms like Twitter. This hate speech has resulted in cyberbullying of certain ethnic groups and targeting of vulnera-ble populations such as the elderly. The authors aim to reduce the Frequency of hate speech usage by identifying hate-related phrases connected to COVID-19 in offensive tweets on Twitter. To achieve this goal, they con-struct a dataset of tweets targeting the elderly population and complement it with another dataset aimed at the Asian population. They evaluate the data using BERT, a self-attention model and identify 186 unique keywords optimized for the Asian population and 100 keywords targeting elderly individuals. Based on their findings, the authors suggest a control mecha-nism that would prompt users to reconsider using specific sensitive phrases detected by their method. They also explore the BERT attention mechanism and discover that the model learns the highest-impact, wide-range attentions at the beginning or end of the layers; it relies on the shape of the basic information. The authors' findings suggest that the BERT model makes predictions using a hate term and a specific group or individual, which aligns with existing hate speech research that shows hate speech is often directed at specific people or groups.

The focus of this study [35] is to scrutinize the occurrence and behavioral consequences of hostile speech in college environments, particularly online. The authors emphasize the harmful consequences of hostile speech and the challenges involved in managing it in university settings. To assess the frequency of hostile speech, the College Hate Index (CHX) is used to evaluate a corpus of 6 million Reddit comments posted in 174 college groups. We will analyze hateful language in different categories and apply causal inference techniques to understand its impact approach to explore the neurological impact of hostile speech on people's online stress manifestation. They analyze language, discriminatory term usage, and personality factors to evaluate individuals' psychological tolerance to hostile speech. The study findings reveal that hostile discourse is prevalent in college subreddits and that exposure to hateful speech increases stress manifestation. The study also identifies that people's psychological endurance to hostile speech differs, with some exhibiting lower endurance than others, and those with lower endurance being more sensitive. Individuals who exhibit lower endurance when compared to individuals with better endurance are more prone to emotional outbursts and neurotic behavior. The study has practical implications and intervention efforts to combat the harmful consequences of online hateful speech in universities, as well as insights into the ethical issues of investigating the impact of hate speech on mental health in online communities. Overall, the research lays the foundation for future investigations into the psychological effects of hostile speech in online communities, particularly those with both online and physical equivalents.

This work [36] focuses on hate speech identification in multimodal publications that include both text and picture. The authors gather and annotate a huge dataset from Twitter, MMHS150K, and present multiple hate speech detection techniques that examine both textual and visual information. They present quantitative and qualitative results by comparing the performance of these multimodal models to that of unimodal detection. Although visuals might be beneficial for hate speech identification, the authors discover that existing multimodal models are not as successful as algorithms that solely examine text. They highlight the task's obstacles and limits and leave the field and dataset open for further investigation. Overall, our research contributes to the development of more complete hate speech detection algorithms that take both textual and visual components of multimodal publications into consideration.

In this study [37], the authors explore the harmful effects of hate speech on the impact of hate speech on social relations and political views. They propose a model to explain the psychological mechanisms underlying these effects. The study suggests that hate speech can lead to detrimental impacts on emotional, behavioral, and normative levels, leading to the replacement of empathy with intergroup disdain. The use of hate speech can establish a descriptive norm that supports derogatory attitudes toward outgroups while also undermining existing anti-discriminatory norms. The study also suggests that desensitization to hate speech can diminish people's capacity to identify the hurtful nature of such language. The authors provide an agent-based modeling technique to investigate the behavior of this model and show that hate speech has the capacity to destroy systems that are successful in limiting its spread. The relevance of understanding the psychological dynamics of hate speech and their influence on intergroup interactions and political radicalization is emphasized in the article.

In this study, the authors focus on evaluating fairness in document classification algorithms, with a particular emphasis on hate speech detection. They argue that previous research has been limited by the use of synthetic monolingual data without author demographic information, which limits the generalizability of results. To address this limitation, the authors create a multilingual Twitter corpus of offensive language identification that includes surmised author demographics such as age, nationality, gender, and race/ethnicity. The corpus includes data from five different languages. To assess the accuracy and fairness of the inferred demographic classifications, the authors use a crowdsourcing platform called Figure Eight. In addition, the authors analyze potential sources of bias by conducting a fact-based analysis of demographic prediction on the English corpus. Finally, the authors evaluate the effectiveness and fairness of four document classifiers and compare them to baseline classifiers based on author-level demographic variables. This study provides valuable insights into the fairness and potential sources of bias evaluation in vitriol detection algorithms, as well as ways to mitigate such bias through more representative datasets and evaluation metrics. Overall, this article adds to the expanding body of knowledge on evaluating fairness in document categorization methods. The authors provide a great resource for future study in this area by compiling and distributing a multilingual Twitter corpus with inferred author demographic variables. Furthermore, their empirical study of demographic prediction and bias might help to design more accurate and fair document classifiers for hate speech detection and other applications.

It is critical to accurately identify hate speech in order to create a safe and inclusive online environment. Researchers assessed the accuracy of eight alternative anti-hate speech methods for categorization: SVM, LR, DT, CNN, Multi-layer Perceptron (MLP), Bidirectional Encoder Representations from

Table 11.1 Literature review

S.No	Language	Platform	Algorithms	Results
1 [3]	English	Social Media	Linear SVM	78% (Accuracy)
2 [4]	English	Twitter	LR NB Linear SVM	95.6%
3 [5]	English	Twitter	DT SVM LR	LR-76% DT-72% SVM-75%
4 [6]	English	Twitter	SVM MLP CNN	CNN-78% MLP-71 SVM-69
5 [7]	Urdu	Twitter	BERT XLM-RoBERTa DistilBERT	F1: 0.68 F1: 0.68 F1: 0.69
6 [8]	Fidel Script	Facebook	CNN	F-Score-92.5
7 [9]	English	Twitter	BiLSTM with Deep CNN & Hierarchical Attention-based DL (BiCHAT)	Accuracy (Validation) BiCHAT-97% Accuracy (Training) BiCHAT-98%
8 [10]	Arab	Arab Social Media	Hybrid (CNN-LSTM)	F1: 82.3%
9 [11]	English	All Social Media	Bi-LSTM, Bi-GRU	ROC-AUC-98.63
10 [12]	English (Hater Net's Public data)	Twitter	Hater BERT, Random Forest	F1–99%
11 [13]	English Social Media Posts	Twitter	NA	NA
12 [14]	Social Media	Twitter	SVM, LR, RF, CNN	78, 78, 76, 89 (Accuracy)
13 [15]	English	Twitter	Fuzzy Logic	F1 Score: 92%
14 [16]	English	Twitter	Genric Hate Speech sentiment classification	95 Accuracy 93 Precision 92 Recall 93 F1-Score
15 [17]	English	Social Media	Deep Multi-task learning (MTL) Framework	NA
16	Misogyny	Twitter	Automatic misogyny identification (LSTM, BERT)	Accuracy: 0.606 Precision: 0.65 F-Score: 0.706 Recall: 0.933

(Continued)

Table 11.1 (Continued) Literature review

S.No	Language	Platform	Algorithms	Results
17 [18]	English	Twitter	CNN-RNN Deep Model ("ABCDM")	Recall Pos-0.95 Neg-0.81 Precision Pos-0.95 Neg-0.83 FI Score Pos-0.95 Neg-0.83 Accuracy 0.92
18 [19]	English	Twitter	SVM RF DT NB XgBoost	Precision • 80.1 • 81.0 • 70.3 • 72.3 • 80.6 Recall • 75.4 • 75.5 • 69.7 • 70.5 • 76.9 FI Score • 75.9 • 75.8 • 69.3 • 69.2 • 77.1

Transformers (BiCHAT), RF, and LSTM networks. Figure 11.1 displays the accuracy results of these methods, with BiCHAT performing the best with an accuracy score of 98, and LSTM performing the worst with a score of 60. The other methods achieved accuracy scores ranging from 71 to 78. These results highlight the potential of the BiCHAT method in hate speech detection and the need for further development of other methods to improve their accuracy.

It is crucial to keep in mind that accuracy is only one of several metrics utilized to assess the performance of hate speech classification models. Other metrics, including precision, recall, and F1-score, can provide a more comprehensive evaluation of the model's performance. Furthermore, the results shown in Figure 1 must be interpreted in light of the particular dataset and evaluation protocol used.

In Figure 11.2, a bar graph presents the F1-score results of various models for hate speech classification, including BERT, XLM-ROBERTa, DistilBERT, CNN, CNN-LSTM, RF, Fuzzy Logic, LSTM, CNN-RNN, SVM, DT, NB, and XgBOOST. It is worth noting that the evaluation

ACCURACY

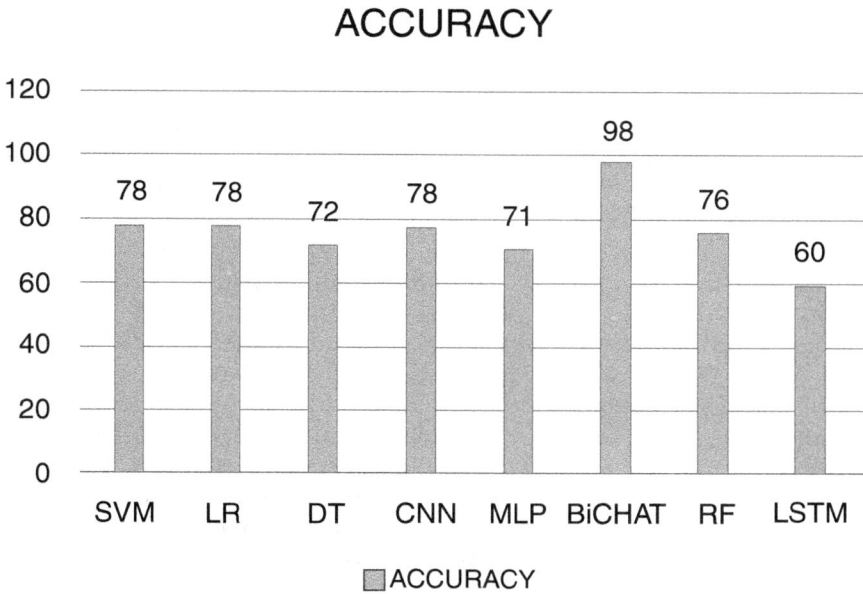

Figure 11.1 Accuracy of various hate speech classification methods.

F1 Score

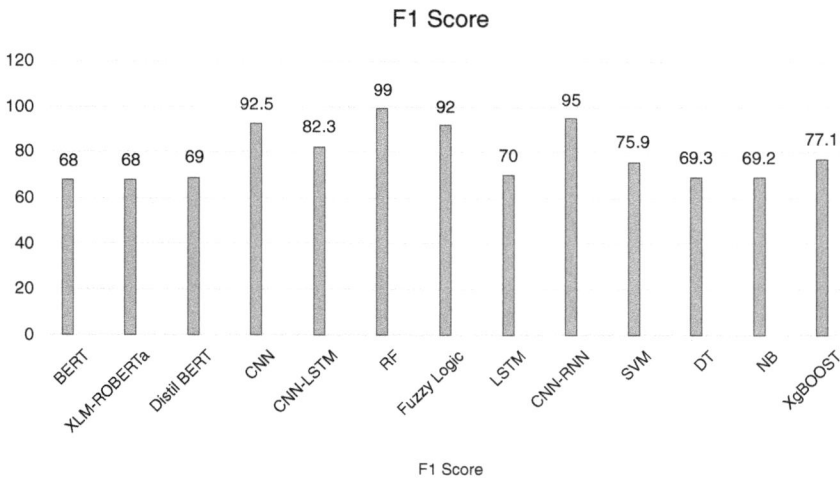

Figure 11.2 Different models' F1 scores for hating speech classification.

metrics used to measure performance vary between studies, and the results should be considered in the context of the specific dataset and evaluation protocol used. The F1-score, which takes precision and recall into account, gives a more thorough evaluation of the model's performance.

According to the data shown in Figure 11.2, the Random Forest model had the greatest F1-score of 99, while the BERT and XLM-ROBERTa models had the lowest F1-score of 68. The other models displayed varying results with F1-scores ranging from 69.2 to 92.5. It is crucial to highlight, however, that the F1-score is only one of the metrics used to assess the efficacy of hate speech categorization algorithms. Other measures, such as accuracy and recall, can offer a more thorough assessment of the strategy performance. Furthermore, the results presented in Figure 11.2 should be interpreted in the context of the specific dataset and evaluation protocol used.

In the domain of hate speech classification, precision and recall are crucial metrics for evaluating the performance of different models. Figure 11.3 presents the precision and recall results of several models, including Sentiment Classification, LSTM, CNN-RNN, SVM, RF, DT, NB, and XgBOOST, shown as a bar graph with two bars for each model. The Sentiment Classification model achieved the highest precision and recall scores of 93 and 92, respectively, while the LSTM model had the highest recall score of 93, but with a relatively low precision score of 65. The other models displayed varying results, with precision and recall scores ranging from 65 to 95. It is important to note that accuracy and recall are not the only measures used to assess the effectiveness of hate speech categorization algorithms. Accuracy and F1-score are also important metrics that can provide a comprehensive evaluation of a model's performance. In addition, the results in Figure 11.3 should be interpreted within the context of the dataset and evaluation protocol utilized.

PRECISION & RECALL

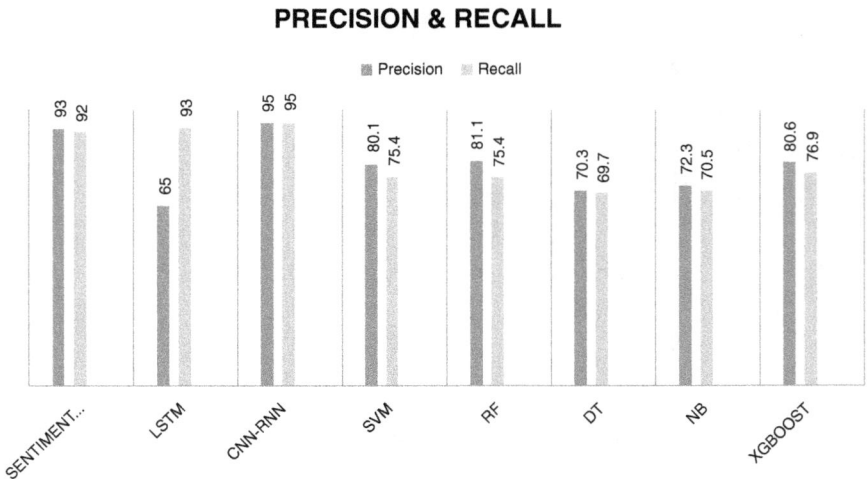

Figure 11.3 Precision and recall of several models for classifying hate speech.

DATASET

Hate speech detection is an important task in NLP and ML, as it can help identify and prevent online harassment, hate crimes, and discrimination. Various datasets have been developed to train ML models to detect hate speech in text. These datasets typically consist of a large number of examples of text labeled as hate speech, offensive speech, or neutral language. The text may be collected from social media platforms, news articles, or other sources. The labeling process is typically performed by human annotators who read each example and assign it a label based on predefined criteria. However, the labeling process can be subjective and prone to biases, which can impact the accuracy and generalizability of the resulting models. As such, it is important to carefully select and evaluate datasets for hate speech detection to ensure they are representative and unbiased. Hate speech detection is a challenging task due to the complexity and subjectivity of identifying hate speech in text. As a result, the development of reliable and diverse datasets is crucial for training and evaluating ML models for this task. To build a dataset for hate speech detection, researchers typically collect large amounts of text data from various sources, including social media platforms, news articles, and online forums. The data may be filtered to exclude irrelevant content, such as spam or non-English text. The remaining text is then labeled by human annotators who are trained to identify instances of hate speech, offensive language, and neutral language. The labeling process may involve the use of guidelines or instructions to ensure consistent and accurate labeling. It is important to note that the labeling process can be challenging, as hate speech can be difficult to define and identify. Different annotators may have different interpretations of what constitutes hate speech, which can lead to inconsistencies in labeling. In addition, the use of automated methods for labeling hate speech may not always be accurate, as they may not account for nuances in language use and context. Therefore, it is important to carefully evaluate and validate hate speech detection datasets to ensure they are reliable and representative. This includes assessing the quality and consistency of labeling, the diversity and representativeness of the data, and the potential biases or limitations of the dataset. Furthermore, researchers should aim to create datasets that are large and diverse enough to support the training of robust and generalizable ML models for hate speech detection.

- The Hate Speech and Offensive Language (OLID) dataset is a popular dataset for hate speech detection, consisting of over 14,000 tweets annotated as hateful, offensive, or neither. The dataset was collected using Twitter's streaming API and was annotated by three human annotators for each tweet. Each tweet in the dataset is labeled as one of three categories: hate speech, offensive language, or neither. Hate speech is defined as language that is intended to degrade, dehumanize, or demean a person or group based on their identity or characteristics. Offensive language is defined as language that is inappropriate or offensive, but not necessarily intended to harm or attack a person or group.

The OLID dataset is widely used for research in hate speech detection and is a popular benchmark for evaluating ML models. The dataset is notable for its large size and the use of multiple human annotators, which helps to reduce the impact of individual biases and inconsistencies in labeling. However, the dataset has been criticized for its limited diversity and potential biases in the labeling process, as it focuses primarily on English-language tweets from a specific platform (Twitter). As with all hate speech detection datasets, careful consideration must be given to their strengths and limitations when using them for research or practical applications.

- The Twitter Hate Speech Dataset is another popular dataset for hate speech detection, consisting of over 24,000 tweets collected using Twitter's search API and annotated by human annotators. The tweets in the dataset are labeled as one of three categories: hate speech, offensive language, or neither. The dataset was created to support research in hate speech detection and is widely used as a benchmark for evaluating ML models. Like the OLID dataset, the Twitter Hate Speech Dataset is notable for its large size and the use of human annotators. However, it has also been criticized for potential biases in the labeling process, such as the over-representation of certain groups or topics. In addition, the use of Twitter as a source of data may limit the generalizability of models trained on this dataset, as hate speech may be expressed differently on other platforms or in different contexts. Despite these limitations, the Twitter Hate Speech Dataset remains a valuable resource for researchers and practitioners working on hate speech detection. Its large size and multiple labels make it well-suited for training and evaluating ML models, while its use of human annotators helps to ensure the quality and consistency of labeling.
- The Wikipedia Detox dataset is a collection of over 100,000 comments from Wikipedia that have been labeled as toxic, severe toxic, obscene, threat, insult, or identity hate. The comments were manually annotated by human annotators to provide a labeled dataset for ML models to use for training and evaluation. The dataset is widely used in NLP research to develop models that can identify and classify toxic or harmful content online.
- The Davidson Hate Speech and Offensive Language dataset is another popular dataset used in NLP research. It contains a collection of over 24,000 tweets that have been labeled as hate speech, offensive language, or neither. The dataset was created by researchers at the University of California, Berkeley, and is named after one of the lead authors, Thomas Davidson. The dataset is often used to develop ML models for detecting hate speech and offensive language on social media platforms.

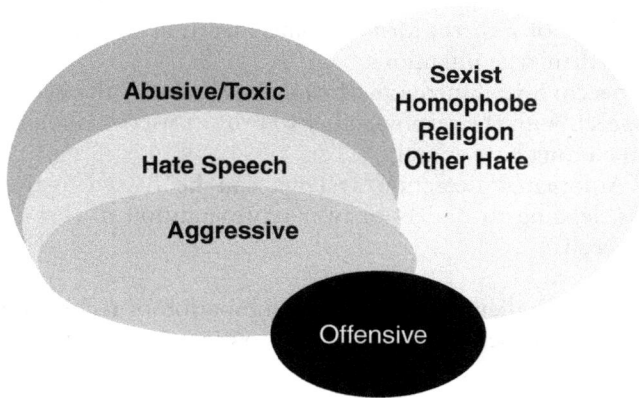

Figure 11.4 Hate speech concept structure.

- The Gab Hate Speech dataset is a collection of 22,484 comments from the social media platform Gab that were collected in 2017 and annotated by human annotators. The dataset was created by researchers at the University of Copenhagen and is often used for developing ML models to detect hate speech on social media platforms. The comments in the dataset have been labeled as either hate speech or not hate speech, and the dataset is notable for its high prevalence of hate speech due to Gab's reputation as a platform that allows hate speech and other forms of extremist content.

These datasets may be used to train ML models to detect hate speech in text automatically. It should be noted, however, that these datasets are not flawless and may contain biases or mistakes in the tagging process. Furthermore, because hate speech is a complicated and subjective term, human annotators may differ on what constitutes hate speech.

Some specific challenges related to hate speech on social media include:

- **Scale:** Social media platforms generate a massive amount of content, making it difficult to manually review and moderate all posts and comments for hate speech.
- **Contextual Understanding:** Detecting hate speech requires understanding the context in which the language is used. Social media posts are often short and may not provide enough context for accurate interpretation.
- **Evolving Language:** Hate speech can take many forms and may use slang, coded language, or memes that are difficult for traditional NLP algorithms to identify.

- **User Behavior:** Some users intentionally try to evade detection by using subtle or indirect forms of hate speech or by using euphemisms to mask their true intentions.
- **Free Speech:** Social media platforms must balance the need to protect free speech with the responsibility to protect users from harmful content, including hate speech.
- **Bias:** Automated detection systems can be biased against certain groups, leading to under- or over-representation of certain types of hate speech.

Addressing these challenges requires a combination of technical solutions, policy changes, and user education efforts to create a safer and more inclusive online environment.

Challenges for ML and DL in the detection of hate speech

There are several difficulties that ML & DL models encounter in recognizing hate speech:

- **Limited Training Data:** Hate speech is a complex and nuanced problem, and collecting labeled training data can be difficult, especially for underrepresented groups.
- **Bias in Training Data:** The effectiveness of ML models is significantly influenced by the quality and representativeness of the training data. Skewed data can lead the model to replicate these biases.
- **Contextual Understanding:** Detecting hate speech requires understanding the context in which language is used. Social media posts are often short and may not provide enough context for accurate interpretation.
- **Evolving Language:** Hate speech can take many forms and may use slang, coded language, or memes that are difficult for traditional NLP algorithms to identify.
- **Generalization:** A model that performs well on a particular dataset or platform may not generalize well to other platforms or languages, leading to poor performance and high error rates.
- **Adversarial Attacks:** Hackers can intentionally manipulate or deceive ML models to evade detection, leading to false negatives or false positives.
- **Trade-Offs between Accuracy and Fairness:** Models optimized for high accuracy may not be fair to all groups, leading to under-representation of certain types of hate speech or over-representation of others.

Addressing these challenges requires ongoing research and development of more robust and accurate ML & DL, as well as careful consideration of the ethical and social implications of hate speech detection algorithms.

DISCUSSION

RQ1: What are the most generally used tactics for identifying hate speech on social media, and how do their accuracy and efficiency compare?

Ans: The most common techniques used for hate speech detection on social media include:

- **NLP-Based Techniques:** Text classification and sentiment analysis are two NLP approaches used to identify hate speech on social media by assessing text and contextual information such as grammar, semantic meaning, and sentiment.
- **ML-Based Techniques:** To identify hate speech on social media, ML algorithms such as decision trees, random forests, and support vector machines are trained on annotated data. To increase accuracy, these strategies can be used with NLP techniques.
- **DL-Based Techniques:** CNN and RNN are DL algorithms that are used to evaluate massive volumes of text data on social media and identify hate speech. In recent investigations, these strategies have yielded positive outcomes.

In terms of accuracy, ML & DL-based techniques tend to perform better than NLP-based techniques, as they can better capture the complexities and nuances of hate speech. However, they can also be more computationally intensive, requiring more computational resources and time to train the algorithms.

In terms of efficiency, NLP-based techniques tend to be faster and require less computational resources, but may not be as accurate as ML and DL-based techniques. The efficiency of ML and DL-based techniques can be improved by using more efficient algorithms, such as lightweight models, or by reducing the size of the training data.

RQ2: What are the various ways in which hate speech is expressed on social media, including but not limited to racism, sexism, and homophobia, and what obstacles are encountered in detecting these forms of hate speech using automated methods?

Ans: Different sorts of hate speech, such as racism, sexism, and homophobia, can emerge in numerous ways on social media. For example:

- **Racism:** Racial slurs, comments promoting racial stereotypes, and racist jokes are all examples of racism on social media. Racism can also be subtler and implicit, making it difficult to detect automatically.
- **Sexism:** Sexist comments and jokes, objectification of women, and promotion of gender stereotypes are examples of sexism on social media. Like racism, sexism can also be implicit and difficult to detect automatically.
- **Homophobia:** Homophobic comments, jokes, and slurs are examples of hate speech on social media that target the LGBTQ+ community.

Detecting these different forms of hate speech automatically is a challenging task, as they can be expressed in various forms and can be subtle or implicit. In addition, different communities and cultures have different sensitivities to hate speech, which can make it difficult to develop a single, universal definition of hate speech. One of the major challenges in detecting hate speech automatically is the need for large amounts of annotated data to train ML algorithms. This data must be diverse and representative of the various forms of hate speech, and it must be annotated accurately to ensure that the algorithms are able to learn the correct patterns. Another challenge is the ability of hate speech to evolve and change over time, making it difficult for algorithms to keep up with new forms of hate speech. For example, hate speech can be expressed using slang or code words, making it difficult for algorithms to detect.

Finally, the accuracy of hate speech detection algorithms can be affected by factors such as the use of irony, sarcasm, and other forms of figurative language, which can make the intent behind a message difficult to interpret. Despite these obstacles, academics and practitioners are working to build and enhance hate speech detection algorithms that employ a combination of NLP, ML, and DL approaches to help alleviate the detrimental consequences of hate speech on social media.

RQ3: What are the limitations of current hate speech detection algorithms, and how can they be improved to better detect and remove hate speech from social media?

Ans: There are several limitations of current hate speech detection algorithms, including:

- **Lack of Variety in Training Data:** One of the biggest shortcomings of existing hate speech detection algorithms is the lack of diversity in the training data used to train them. This can result in biased algorithms that are less effective at detecting hate speech in underrepresented groups or in different languages or cultures.
- **Difficulty in Detecting Implicit and Covert Hate Speech:** Another limitation of current algorithms is their difficulty in detecting implicit and covert hate speech, such as hate speech that is expressed using sarcasm or irony. This type of hate speech can be difficult to detect, as it requires a deeper understanding of the context and intent behind the text.
- **Inadequate Evaluation Metrics:** A third limitation of current hate speech detection algorithms is the use of inadequate evaluation metrics. For example, metrics such as accuracy and F1-score can be misleading, as they do not always reflect the real-world performance of the algorithms.
- **Overreliance on Keyword-Based Methods:** Another limitation of current hate speech detection algorithms is their overreliance on

keyword-based methods, such as lists of hate speech words or expressions. These methods can be easily circumvented by hate speech perpetrators who use euphemisms or other methods to avoid detection.

To overcome these limitations, researchers and practitioners can improve hate speech detection algorithms by:

- **Increasing the Diversity of the Training Data:** To reduce bias in hate speech detection algorithms, it is important to ensure that the training data is diverse and representative of different cultures, languages, and communities.
- **Developing Algorithms That Can Detect Implicit and Covert Hate Speech:** Researchers can develop algorithms that are better equipped to detect implicit and covert hate speech, by incorporating information about the context and intent behind the text.
- **Using More Robust Evaluation Metrics:** Researchers can use more robust evaluation metrics, such as human evaluation or cross-validation, to better evaluate the effectiveness of hate speech detection algorithms. These measures better represent the algorithms' real-world performance.
- **Incorporating Information from Multiple Sources:** To improve the robustness of hate speech detection algorithms, researchers can incorporate information from multiple sources, such as social media platforms, user profiles, and image and video data.

Developing algorithms that can adapt to changing hate speech patterns: To stay ahead of evolving hate speech patterns, researchers can develop algorithms that are capable of adapting to changing hate speech patterns, by using techniques such as transfer learning and active learning.

CONCLUSION

To conclude the research covered in this article suggests that ML and DL algorithms can detect hate speech on social media. The BiCHAT technique had the best accuracy score, while the Random Forest model had the highest F1-score. The Sentiment Classification method scored best in terms of precision and recall. The LSTM approach, on the other hand, had relatively poor accuracy and precision ratings. It is important to keep in mind that these conclusions are based on a small number of research papers and may not be applicable to all social media platforms. More research is needed to solve the present approach limitations and increase the accuracy and efficiency of hate speech detection systems.

This includes increasing the diversity of training data, detecting implicit and covert hate speech, using robust evaluation metrics, incorporating multiple sources, and developing algorithms that can adapt to changing hate speech patterns.

REFERENCES

[1] P. William, R. Gade, R. Chaudhari, A. B. Pawar, and M. A. Jawale, "ML based Automatic Hate Speech Recognition System," in *2022 International Conference on Sustainable Computing and Data Communication Systems (ICSCDS)*, Apr. 2022, pp. 315–318. doi: 10.1109/ICSCDS53736.2022.9760959.

[2] H. Simon, B. Baha, and E. Garba, "Trends in MLON Automatic Detection of Hate Speech on Social Media Platforms: A Systematic Review," vol. 7, pp. 1–16, May 2022.

[3] "Unintended bias evaluation: An analysis of hate speech detection and gender bias mitigation on social media using ensemble learning | Elsevier Enhanced Reader." https://reader.elsevier.com/reader/sd/pii/S095741742200447X?token=F26FF 773EB6D6184940993893AA1D4CC56BDB3B4B012498C0EF9F0F2BCAA48 CC88A9CBB21EECACD0A8D289E0CBFAEC1A&originRegion=eu-west-1& originCreation=20230208035003 (accessed Feb. 08, 2023).

[4] J. S. Malik, G. Pang, and A. van den Hengel, "DL for hate speech detection: A comparative study." arXiv, Feb. 18, 2022. Accessed: Feb. 08, 2023 [Online]. Available: https://arxiv.org/abs/2202.09517

[5] "Hate speech detection on Twitter using transfer learning." Elsevier Enhanced Reader. https://reader.elsevier.com/reader/sd/pii/S0885230822000110?token= 943DEE98D9C71AEC5368EFB5CAA77ABCA941ABD5F90490D04BBD466E 205DDECC9BD6172839440C736D07FE73D6848E91&originRegion=eu-west-1& originCreation=20230208035755 (accessed Feb. 08, 2023).

[6] S. Khan, M. Fazil, V. K. Sejwal and M.A. Alshara"BiCHAT: BiLSTM with deep CNN and hierarchical attention for hate speech detection." Journal of King Saud University - Computer and Information Sciences 34(4), Elsevier Enhanced Reader. DOI:10.1016/j.jksuci.2022.05.006

[7] H. Elzayady, M. Mohamed, K. Badran, and G. Salama, "A hybrid approach based on personality traits for hate speech detection in arabic social media," pp. 1979–1988, 2022. doi: 10.11591/ijece.v13i2.

[8] A. Chhabra and D. K. Vishwakarma, "A literature survey on multimodal and multilingual automatic hate speech identification," *Multimed. Syst.*, Jan. 2023, doi: 10.1007/s00530-023-01051-8.

[9] A. C. Mazari, N. Boudoukhani, and A. Djeffal, "BERT-based ensemble learning for multi-aspect hate speech detection," *Clust. Comput.*, Jan. 2023, doi: 10.1007/s10586-022-03956-x.

[10] "A curated dataset for hate speech detection on social media text." Elsevier Enhanced Reader. https://reader.elsevier.com/reader/sd/pii/S23523409220 10356?token=E96896C2DDAFBB424BDADD55D6FC69719658CF24F90 5C3E4E9D1FB14638D4259DF3ADE3A59C477A527D78586FE5CFBD9& originRegion=eu-west-1&originCreation=20230208043413 (accessed Feb. 08, 2023).

[11] G. del Valle-Cano, L. Quijano-Sánchez, F. Liberatore, and J. Gómez, "SocialHaterBERT: A dichotomous approach for automatically detecting hate speech on Twitter through textual analysis and user profiles," *Expert Syst. Appl.*, vol. 216, p. 119446, Apr. 2023, doi: 10.1016/j.eswa.2022.119446.

[12] K. Solovev and N. Pröllochs, "Moralized language predicts hate speech on social media," *PNAS Nexus*, vol. 2, no. 1, p. pgac281, Jan. 2023, doi: 10.1093/pnasnexus/pgac281.

[13] M. A. Ullah, S. M. Marium, S. A. Begum, and N. S. Dipa, "An algorithm and method for sentiment analysis using the text and emoticon," *ICT Express*, vol. 6, no. 4, pp. 357–360, Dec. 2020, doi: 10.1016/j.icte.2020.07.003.

[14] F. E. Ayo, O. Folorunso, F. T. Ibharalu, I. A. Osinuga, and A. Abayomi-Alli, "A probabilistic clustering model for hate speech classification in Twitter," *Expert Syst. Appl.*, vol. 173, p. 114762, Jul. 2021, doi: 10.1016/j.eswa.2021. 114762.

[15] F. E. Ayo, O. Folorunso, F. T. Ibharalu, and I. A. Osinuga, "MLtechniques for hate speech classification of twitter data: State-of-the-art, future challenges and research directions," *Comput. Sci. Rev.*, vol. 38, p. 100311, Nov. 2020, doi: 10.1016/j.cosrev.2020.100311.

[16] P. Kapil and A. Ekbal, "A deep neural network based multi-task learning approach to hate speech detection," *Knowl. Based Syst.*, vol. 210, p. 106458, Dec. 2020, doi: 10.1016/j.knosys.2020.106458.

[17] E. W. Pamungkas, V. Basile, and V. Patti, "Misogyny detection in Twitter: A multilingual and cross-domain study," *Inf. Process. Manag.*, vol. 57, no. 6, p. 102360, Nov. 2020, doi: 10.1016/j.ipm.2020.102360.

[18] M. E. Basiri, S. Nemati, M. Abdar, E. Cambria, and U. R. Acharya, "ABCDM: An attention-based bidirectional CNN-RNN deep model for sentiment analysis," *Future Gener. Comput. Syst.*, vol. 115, pp. 279–294, Feb. 2021, doi: 10.1016/j.future.2020.08.005.

[19] P. Chitra et al., "Sentiment analysis of product feedback using natural language processing," *Mater. Today Proc.*, Feb. 2021, doi: 10.1016/j. matpr.2020.12.1061.

[20] A. Arango, J. Pérez, and B. Poblete, "Hate speech detection is not as easy as you may think: A closer look at model validation (extended version)," *Inf. Syst.*, vol. 105, p. 101584, Mar. 2022, doi: 10.1016/j.is.2020.101584.

[21] S. Kaur, S. Singh, and S. Kaushal, "Abusive content detection in online user-generated data: A survey," *Procedia Comput. Sci.*, vol. 189, pp. 274–281, Jan. 2021, doi: 10.1016/j.procs.2021.05.098.

[22] G. Kovács, P. Alonso, and R. Saini, "Challenges of hate speech detection in social media," *SN Comput. Sci.*, vol. 2, no. 2, p. 95, Feb. 2021, doi: 10.1007/s42979-021-00457-3.

[23] M. A. Ullah et al. "An algorithm and method for sentiment analysis using the text and emoticon," *ICT Express*, vol. 6, no. 4, pp. 357–360, 2020.

[24] P. Kapil and A. Ekbal, "A deep neural network based multi-task learning approach to hate speech detection," *Knowl. Based Syst.*, vol. 210, p. 106458, 2020.

[25] A. Matamoros-Fernández and J. Farkas, "Racism, hate speech, and social media: A systematic review and critique," *Television New Media*, vol. 22, no. 2, pp. 205–224, 2021.

[26] A. Al-Hassan and H. Al-Dossari, "Detection of hate speech in social networks: A survey on multilingual corpus," in *6th International Conference on Computer Science and Information Technology*, vol. 10, 2019.

[27] Z. Mossie and J.-H. Wang, "Vulnerable community identification using hate speech detection on social media," *Inform. Process. Manag.*, vol. 57, no. 3, p. 102087, 2020.

[28] B. Mathew et al. "Spread of hate speech in online social media," in *Proceedings of the 10th ACM Conference on Web Science*, 2019.

[29] M. Mozafari, R. Farahbakhsh, and N. Crespi, "A BERT-based transfer learning approach for hate speech detection in online social media," in *Complex Networks and Their Applications VIII: Volume 1 Proceedings of the Eighth International Conference on Complex Networks and Their Applications COMPLEX NETWORKS 2019 8*, Springer International Publishing, 2020.

[30] M. Corazza et al. "A multilingual evaluation for online hate speech detection," *ACM Trans. Internet Tech.*, vol. 20, no. 2, pp. 1–22, 2020.

[31] B. He et al. "Racism is a virus: Anti-Asian hate and counterspeech in social media during the COVID-19 crisis," in *Proceedings of the 2021 IEEE/ACM International Conference on Advances in Social Networks Analysis and Mining*, 2021.

[32] R. Rogers, "Deplatforming: Following extreme Internet celebrities to Telegram and alternative social media," *Eur. J. Commun.*, vol. 35, no. 3, pp. 213–229, 2020.

[33] N. Vishwamitra et al. "On analyzing covid-19-related hate speech using Bert attention," in *2020 19th IEEE International Conference on ML and Applications (ICMLA)*, IEEE, 2020.

[34] K. Saha, E. Chandrasekharan, and M. De Choudhury, "Prevalence and psychological effects of hateful speech in online college communities," in *Proceedings of the 10th ACM Conference on Web Science*, 2019.

[35] R. Gomez et al. "Exploring hate speech detection in multimodal publications," in *Proceedings of the IEEE/CVF Winter Conference on Applications of Computer Vision*, 2020.

[36] M. Bilewicz and W. Soral. "Hate speech epidemic. The dynamic effects of derogatory language on intergroup relations and political radicalization," *Political Psychol.*, vol. 41, pp. 3–33, 2020.

[37] X. Huang et al. "Multilingual twitter corpus and baselines for evaluating demographic bias in hate speech recognition." *arXiv preprint arXiv:2002.10361*, 2020.

Chapter 12

Digital twin technology for industrial automation

Anil Kumar Dubey and Richa Bansal

OVERVIEW

In essence, a digital twin [1] is a computer software that simulates how a process or product would work using data from the real world. The simulation is based on both historical data and the asset's present state. Virtual copies of actual objects, procedures, or systems are known as digital twins. Before producing prototypes or installing equipment, they can be used in industrial automation to:

- Design and commission systems in the virtual environment.
- Produce digital replicas of actual machines.
- Merge digital and physical data to comprehend the functionality and potential worth of physical equipment.
- Develop along with automation itself.

A crucial piece of the technology of Industry 4.0, or the Fourth Industrial Revolution, is digital twins. They build simulations that can forecast a product's or process' performance using real-world data. The simulation is based on both the asset's present state and previous data. With the use of digital twins, it is possible to increase system performance while keeping costs low, decrease physical testing, and enhance product quality.

An effective instrument in the realm of industrial automation [2] is digital twin technology. It describes the process of creating a digital representation of a physical system or procedure. The physical system can be observed, examined, simulated, and controlled using this digital twin. The usage of digital twin technology in industrial automation has a number of advantages.

- **Real-Time Monitoring:** Real-time information about the state of a physical system is provided by digital twins. Operators and engineers can keep track of the system's performance and health by integrating sensors and data streams from the physical equipment into the digital twin.

DOI: 10.1201/9781003479031-12

- **Predictive Maintenance:** It is feasible to foresee when repair is required by continuously monitoring the equipment through digital twins. This can avoid unplanned downtime and lower maintenance expenses.
- **Process Optimization:** Digital twins make it possible to simulate different outcomes and do "what-if" assessments. They can be used by engineers to streamline procedures and make knowledgeable selections regarding alterations or advancements.
- **Quality Control:** Digital twins can be used in manufacturing to track product quality. Deviations can be found and fixed by comparing the digital twin's specs with real-time data from the production line.
- **Remote Control:** Digital twins can be used to automate [16] and remotely control industrial processes. Adjustments made by operators to the digital twin are reflected in the physical system.
- **Training and Simulation:** For technicians who do maintenance and operations, digital twins can be used as training aids. In a secure and controlled virtual environment, they can practice equipment operation and troubleshooting.
- **Energy Efficiency:** By evaluating data and suggesting strategies to cut energy use without sacrificing productivity, digital twins can assist in minimizing energy consumption in industrial processes.
- **Supply Chain Optimization:** Digital twins can be used to expand beyond specific factories or processes across the entire supply chain. As a result, the entire industrial ecosystem can better coordinate and make decisions.
- **Lifecycle Management:** A product or asset's whole lifecycle can be covered by digital twins, from design and prototyping to production, use, and end-of-life considerations. Better product design and sustainable practices may result from this.
- **Safety and Compliance:** Without endangering people or equipment, digital twins can be used to verify safety procedures and simulate emergency circumstances. They also aid in making sure that regulations are followed.
- **Data Analytics and AI:** AI and sophisticated analytics tools can be used to examine the data gathered from digital twins. This can reveal obscure information and trends that result in additional optimizations.
- **Collaboration:** By offering a shared digital representation of the system, digital twins can improve communication across various teams and departments within an organization as well as with outside partners.

Organizations must make investments in sensor technology, data integration, reliable simulation software, and data analytics skills in order to successfully utilize digital twins for industrial automation. Additionally, it is essential to think about data security and privacy issues because access to digital twin data needs to be restricted to avoid unauthorized alteration or data breaches.

HISTORY

Over several decades, the idea of using digital twin technology for industrial automation [4, 10] has developed. Its roots are found in a number of industries, including manufacturing, aircraft, and simulation. Here is a brief overview of how digital twin technology has developed and changed throughout time.

- **1950s–1960s**: Birth of computer-aided manufacturing (CAM) and computer-aided design (CAD): The development of CAD and CAM technologies in the 1950s and 1960s can be credited for inspiring the development of digital twin technology. These early systems allowed engineers and designers to simulate manufacturing processes and produce digital representations of real-world things. Digital twin technology has its roots in the early days of computer modeling and simulation. Computers were first used by engineers and scientists to model physical systems mathematically, allowing them to simulate and study real-world occurrences.
- **1960s–1970s**: Computer Numerical Control (CNC) machine introduction: The manufacturing sector underwent tremendous breakthroughs in the 1960s with the advent of CNC machines. These devices were early examples of digital control and automation because they used computer programs to manage the milling tools.
- **1970s–1980s**: Aerospace simulations: Digital twin concepts [3] for complex systems were pioneered by NASA and other aerospace agencies. To simulate and examine mission scenarios, structural integrity, and system performance, they constructed digital models of spacecraft and airplanes. Complex system modeling and aerospace: The aerospace sector was one of the first to adopt digital twin ideas. To simulate and analyze space operations, NASA and other companies started creating complex system models that represented complete spacecraft. These early digital twins were utilized for system validation, failure analysis, and mission planning.
- **1990s**: Manufacturing expansion: The 1990s saw the introduction of digital twin concepts. Digital twins have been used by businesses like General Motors and Boeing to imitate manufacturing processes, increase output, and enhance product quality. Early versions of these apps mostly dealt with digital mockups and product lifecycle management (PLM). Rise of Virtual Reality and 3D Modeling: Virtual reality and 3D modeling technologies came into being in the 1990s. These developments made it possible for digital representations of real-world items and places to be more immersive and interactive, establishing the framework for sophisticated digital twins.

- **Early 2000s:** Adoption in manufacturing: Organizations like General Motors and Boeing started implementing digital twin concepts more widely in manufacturing operations. Virtual prototypes and simulations were made using digital twins to evaluate product ideas and manufacturing procedures.
- **Mid-2000s:** Internet of Things (IoT) and smart systems: The middle of the 2000s saw a shift toward IoT and smart systems. As sensors and connectivity proliferated, businesses began incorporating real-time data collection from physical assets into digital models.
- **2000s:** Cyber-physical system (CPS) development: The 2000s saw a rise in popularity for the idea of cyber-physical systems. These technologies created digital representations and control systems using physical components, laying the groundwork for more dynamic and integrated digital twins.
- **Mid-2010s:** CPS and Industry 4.0: The idea of Industry 4.0, which emphasizes the incorporation of digital technology into industrial processes, has gained traction. CPS, which link physical systems with digital representations to create more dynamic digital twins, have emerged as a crucial element. Integrating IoT with Big Data: The development of digital twin technology was considerably expedited by the 2010s IoT and Big Data technology convergence. Real-time data collection from physical assets thanks to IoT sensors and data streams allowed for the creation of more precise and dynamic digital twins.
- **2010s:** Integration of IoT and cloud computing: The fusion of IoT and cloud computing enabled businesses to gather, store, and analyze enormous amounts of real-time data. As a result, it was able to develop digital twins that were incredibly accurate and quick to respond, providing remote monitoring, process optimization, and predictive maintenance. Industry 4.0 and smart production: The Fourth Industrial Revolution, often known as Industry 4.0, placed a strong emphasis on the automation of production processes and the integration of digital technology. To enable predictive maintenance, process optimization, and data-driven decision-making, digital twins became essential to the Industry 4.0 vision.
- **2020s–Present:** Digital twin technology has developed quickly in recent years, and it has been widely adopted. To increase productivity, decrease downtime, and improve product quality, it is used in a variety of industries, including manufacturing, energy, healthcare, and transportation. With the rise of cloud-based digital twin platforms, the technology has also become more widely available. Expanded adoption and ongoing developments: The development of digital twin technology has accelerated recently. A wide range of industries, including energy, healthcare, agriculture, and smart cities, are now using it. The capabilities of digital twins have been further increased by developments in machine learning and artificial intelligence [11].

Modern industrial automation relies heavily on digital twin technology [9], which creates a digital counterpart of real systems and assets. It enables real-time monitoring, preventive maintenance, modeling, and optimization, assisting enterprises in becoming more efficient and competitive in a technological environment that is continually changing. Applications for the technology are anticipated to increase as it develops. The importance of digital twin technology in industrial automation is growing as it develops. Its uses have far surpassed the first prototypes and simulations, and are now crucial for real-time monitoring, predictive analytics, and decision support in intricate industrial systems and processes. Future developments in digital twin technology promise progressively more complex and networked digital duplicates of real-world objects and settings.

EXAMPLE

Digital twin for predictive maintenance in a manufacturing plant

Imagine a facility that makes components for automobiles. The factory uses digital twin technology for preventive maintenance to ensure continuous operation and minimize downtime.

- **Data Collection:** All crucial pieces of gear and equipment throughout the production line have sensors fitted. These sensors gather information about characteristics including temperature, vibration, pressure, and energy usage continually. The centralized system receives this data after that.
- **Digital Twin Creation:** A digital twin is built for each piece of machinery in the production plant using this real-time data. The great level of detail in these digital twins includes 3D models of the machinery, historical information, and performance parameters.
- **Real-Time Monitoring:** The real-time status of each piece of equipment in the plant is shown on a dashboard that is accessible to operators and maintenance staff. They can determine if any equipment is operating abnormally.
- **Predictive Analytics:** The information from the digital twins is processed using machine learning techniques. These algorithms can spot irregularities and patterns in the data. For instance, they are able to pick up on minute variations in vibration patterns that might be signs of a worn-out bearing.
- **Predictive Maintenance Alerts:** The algorithms notify maintenance staff when they find a potential problem. This alert contains details about the potential issue, its seriousness, and suggested solutions.

- **Maintenance Planning:** The digital twin data can be used by maintenance teams to better plan their treatments. Instead of responding to unanticipated failures, they can arrange repair during planned downtime, minimizing production disruptions.
- **Simulation and Testing:** The digital twin can be used by maintenance crews to model repairs prior to completing maintenance. They may simulate various scenarios and order the required components beforehand, ensuring a seamless and effective repair procedure.
- **Historical Analysis:** Engineers are able to evaluate the equipment's long-term performance thanks to the historical data that is recorded in the digital twins. They can spot patterns and decide on equipment replacement or improvements with knowledge.
- **Energy Efficiency:** Additionally, digital twins aid in energy efficiency. The plant may discover places where energy is being wasted and take corrective action to save energy expenses by evaluating energy consumption data from the digital twins.
- **Continuous Improvement:** The information gathered from the digital twins is utilized to refine equipment, processes, and maintenance procedures over time, which boosts productivity and lowers costs.

In this illustration, digital twin technology not only aids in preventing unanticipated equipment failures but also boosts overall operational efficiency in the manufacturing [5, 6, 8] facility by streamlining maintenance procedures, decreasing downtime, and reducing unproductive time. This makes it an important tool for industrial automation because it reduces costs and increases production [7].

PROBLEM WITH DIGITAL TWIN TECHNOLOGY

Although digital twin technology has many benefits for industrial automation, there are a number of difficulties and issues with its application.

- **Data Integration:** It can be challenging and time-consuming to combine data from numerous sensors and systems into a coherent digital twin. The compatibility of many data formats, standards, and protocols can be difficult.
- **Data Quality:** It is crucial that the data gathered by sensors is accurate and reliable. Digital twins can be flawed and findings can be wrong as a result of inaccurate or missing data.
- **Scalability:** It can be resource-intensive to manage digital twins for big and complicated industrial systems. It can be difficult to make sure that the digital twin scales efficiently with the size of the operation.
- **Security:** Sensitive information about machinery and industrial operations is contained in digital twins. A top priority is preventing unwanted access and online threats to sensitive data.

- **Privacy:** Data obtained by digital twins might raise privacy concerns, especially when they involve personal information or data about employees.
- **Cost:** Digital twin development and upkeep can be costly. The advantages must outweigh the costs of sensors, data storage, software development, and upkeep.
- **Skill Gap:** Data science, IoT, and simulation technology expertise may be needed for the implementation and management of digital twins. There may be a skills shortage in these areas for many firms.
- **Interoperability:** It is essential to make sure that digital twins can communicate with other systems and technologies in the industrial ecosystem. It can be tough to achieve smooth interoperability.
- **Regulatory Compliance:** Making sure digital twins conform with the tight regulatory requirements imposed on some businesses can be challenging.
- **Complexity:** Particularly for really complicated systems, digital twins can become quite complex. For operators and engineers, managing and comprehending these complex models can be overwhelming.
- **Data Overload:** Data overload can result from the massive amounts of data that sensors and systems generate. Without the right data analytics tools, sorting through this data to get valuable insights can be overwhelming.
- **Resistance to Change:** Because they worry about losing their jobs or don't grasp its advantages, employees and management may be reluctant to use digital twin technology.
- **Ethical Concerns:** When using digital twins to monitor and manage human activities, such as measuring employee productivity, ethical concerns may arise.
- **Environmental Impact:** Data centers, sensors, and other necessary infrastructure for digital twins may have an impact on the environment in terms of energy use and electrical waste.
- **Reliability and Validation:** A constant problem is ensuring that the digital twin appropriately depicts the physical system it replicates. Validation and calibration must be performed often.

Despite these difficulties, a lot of businesses believe that digital twin technology has more advantages than disadvantages. Planning carefully, investing in technology and training, and being dedicated to data security and quality are frequently required to address these issues. Some of these problems might become less significant as the technology develops, but firms using digital twin solutions will still need to take them into account.

CASE STUDY OF DIGITAL TWIN TECHNOLOGY FOR INDUSTRIAL AUTOMATION

Table 12.1 gives two examples of industrial automation applications using digital twin technology.

Table 12.1 Case study of Siemens Gas Turbine Digital Twin and Digital Twin for Automotive Manufacturing

Company	Overview	Key features	Benefits
Siemens Energy (**Siemens Gas Turbine Digital Twin**)	For use in power generation, Siemens Energy created a digital twin for its gas turbines. With the help of the digital twin, these vital assets should function better, have better maintenance procedures, and experience less downtime.	**Real-Time Monitoring:** Sensors mounted on the gas turbines gather information on a variety of factors, including temperature, pressure, and vibration, continually. To generate a digital twin of each turbine, real-time data is employed. **Predictive Maintenance:** Siemens uses predictive analytics to find irregularities in the performance of the turbines. For instance, variations in vibration patterns may be a sign of trouble. These metrics are used to create maintenance notifications. **Simulation and Testing:** Testing and simulation are made possible by the digital twin, which spares the actual turbine from damage during either process. This enhances effectiveness and performance.	**Reduced Downtime:** Siemens has decreased unexpected downtime for its gas turbines by proactively resolving maintenance issues based on predictive analytics, increasing operational efficiency. **Improved Performance:** Improved turbine performance as a result of optimization through the digital twin has led to better energy output and lower fuel usage. **Cost Savings:** Reduced maintenance costs and more revenue from greater energy generation have resulted from predictive maintenance and optimized performance.

(Continued)

Table 12.1 (Continued) Case study of Siemens Gas Turbine Digital Twin and Digital Twin for Automotive Manufacturing

Company	Overview	Key features	Benefits
Ford Motor Company (**Digital Twin for Automotive Manufacturing**)	Ford uses digital twin technology to improve product quality and production efficiency in its manufacturing processes.	**Process Simulation**: Ford builds digital twins of every component of its production lines, including the robots, conveyors, and assembly lines. By simulating the production process, they can find any potential bottlenecks or inefficiencies. **Quality Control**: Real-time product quality monitoring is done using the digital twin. Data regarding each vehicle's construction is captured by cameras and sensors on the assembly line, and the digital twin compares this information to quality requirements. **Maintenance Planning**: Before machines and robots break down, maintenance tasks are scheduled using predictive maintenance. This reduces production process interruptions to a minimum.	**Increased Efficiency**: Ford uses digital twin simulations to improve production line layouts and procedures, resulting in more quickly and effectively produced goods. **Higher Quality**: With real-time quality management, Ford's strict quality standards are met, lowering faults and raising customer satisfaction. **Cost Reduction**: Predictive maintenance saves the organization a lot of money by reducing unplanned downtime and maintenance costs. **Flexibility**: Ford can swiftly adapt production lines for various vehicle types using the digital twin, increasing manufacturing flexibility.

These case studies demonstrate how digital twin technology is being used in various industrial contexts to increase processes, lower costs, and improve overall quality and efficiency. Digital twins have shown to be useful tools in a variety of industries, including the manufacturing of automobiles and energy, and their adoption is anticipated to increase as the technology develops.

ADVANTAGES AND DISADVANTAGES OF DIGITAL TWIN TECHNOLOGY FOR INDUSTRIAL AUTOMATION

In the context of industrial automation, digital twin technology has both benefits [14] and drawbacks. Here is a list of the main benefits and drawbacks.

Advantages

- **Predictive Maintenance:** Predictive maintenance, made possible by digital twins, assists in finding and fixing equipment problems before they result in expensive failures. This lowers maintenance expenses and downtime.
- **Efficient Operations:** Digital twins can increase operational effectiveness, decrease waste, and increase overall production by continuously monitoring and adjusting processes.
- **Real-Time Monitoring:** Real-time monitoring of industrial systems and equipment by operators and engineers enables prompt reaction to anomalies or breakdowns.
- **Risk Mitigation:** With the help of digital twins, enterprises can detect and reduce operational risks by simulating a variety of situations.
- **Improved Product Quality:** By comparing the specifications of the digital model with current production data, digital twins in manufacturing can improve quality control and decrease faults.
- **Energy Efficiency:** By evaluating sensor data and recommending energy-saving strategies, digital twins can optimize energy use.
- **Cost Savings:** Predictive maintenance, streamlined procedures, and reduced downtime lead to cost savings in maintenance, energy, and operational efficiency.
- **Lifecycle Management:** The full lifecycle of products and assets can be covered by digital twins, assisting businesses in making wise choices regarding design, manufacture, use, and upkeep.
- **Simulation and Testing:** With the use of digital twins, testing and experimenting may be done virtually rather than using physical prototypes and tests, which can save time and resources.

Disadvantages

- **Complexity:** Digital twin implementation and management can be challenging and require knowledge of data integration, modeling, and analytics.

- **Data Integration Challenges:** It can be difficult to combine data from different sensors and systems into a coherent digital twin because of problems with data format, standards, and interoperability.
- **Data Quality:** It is essential to guarantee the precision and dependability of data gathered through sensors. Incorrect insights may result from inaccurate data.
- **Costs:** Digital twin creation and upkeep can be expensive due to the need for sensors, data storage, software development, and employee training.
- **Security Concerns:** Digital twins are possible targets for cyberattacks because they hold private information about machinery and industrial operations. Strong cybersecurity [15] precautions are necessary.
- **Scalability:** It can be difficult and resource-intensive to scale digital twin implementations to span large and sophisticated industrial systems.
- **Regulatory Compliance:** Monitoring and adjusting digital twins to ensure they adhere to industry-specific regulations can be difficult.
- **Privacy Issues:** Data obtained by digital twins might raise privacy concerns, especially when they involve personal information or data about employees.
- **Resistance to Change:** Because they are concerned about losing their jobs or don't comprehend its advantages, staff members and management may be reluctant to adopt digital twin technology.
- **Ethical Considerations:** The usage of digital twins can bring up moral concerns about privacy and surveillance, especially when used to monitor and regulate human activity.
- **Reliability and Validation:** It can be difficult to ensure that the digital twin adequately represents the physical system it mimics and constant validation and calibration are needed.

In conclusion, while digital twin technology has many benefits for industrial automation, such as increased effectiveness, cost savings, and preventive maintenance, it also has drawbacks in terms of complexity, data quality, security, and privacy. When introducing digital twins, businesses must carefully take into account these characteristics and take precautions to minimize any potential drawbacks.

APPLICATION OF DIGITAL TWIN TECHNOLOGY FOR INDUSTRIAL AUTOMATION

The use of digital twin technology for automation and optimization is found in many different industrial areas (Table 12.2).

The adaptability and influence of digital twin technology in industrial automation are shown by these applications. Organizations can acquire insights, optimize operations, and make data-driven decisions to increase efficiency, save costs, and improve overall performance by constructing digital duplicates of physical systems and processes.

Table 12.2 Some important ways that industrial automation is using digital twin technology

Manufacturing and Production	**Smart Factories:** With the use of digital twins, entire manufacturing [19] facilities—complete with equipment and production lines—can be digitally recreated. This makes it possible to monitor, improve, and simulate production processes in real time. **Quality Control:** Digital twins are used to track product quality and spot flaws instantly. Alerts for remedial action are sent out for any deviations from the expected standards. **Inventory Management:** By tracking materials and components in real time, digital twins can optimize inventory levels and make sure that production lines have the resources they need to run effectively.
Predictive Maintenance	**Equipment Health Monitoring:** Industrial machinery and equipment are continuously monitored via digital twins. To identify anomalies and anticipate maintenance requirements, predictive analytics is used, which lowers maintenance costs and downtime.
Energy Management	**Energy Efficiency:** By monitoring and analyzing data from sensors to find potential for energy savings, digital twins aid in the optimization of energy use. **Grid Management:** Digital twins can simulate and control power distribution grids in the energy sector, allowing utilities to react to demand changes and outages more quickly.
Supply Chain Management	**End-to-End Visibility:** Digital twins give firms complete visibility into supply chain operations, enabling them to better demand forecasting, trace shipments, and optimize logistics. **Inventory Optimization:** Companies may eliminate surplus stock and improve inventory levels while ensuring on-time deliveries by digitally modeling the supply chain.
Health and Safety	**Worker Safety:** Utilizing digital twins, working settings can be simulated and analyzed to find safety risks and enhance employee safety procedures.
Water and Wastewater Management	**Water Treatment Plants:** Water treatment operations are monitored and optimized with the use of digital twins, resulting in effective and environmentally friendly management of water resources.

(Continued)

Table 12.2 (Continued) Some important ways that industrial automation is using digital twin technology

Building Management	**Smart Buildings:** Building HVAC, lighting, and security systems are all monitored and controlled by digital twins of the structures. As a result, residents experience increased comfort and energy savings.
Agriculture	**Precision Agriculture:** Digital twins in agriculture enable more precise resource allocation and higher yields by monitoring soil conditions, weather, and crop health.
Aerospace and Defense	**Aircraft Maintenance:** Predictive maintenance in the aerospace sector uses digital twins of aircraft engines and other parts to increase the safety and dependability of the aircraft.
Mining and Natural Resources	**Mining Operations:** By keeping an eye on the machinery, checking the grades of the ore, and enhancing safety procedures, digital twins can optimize mining operations.
Transportation and Logistics	**Fleet Management:** Digital twins are used to track and improve the efficiency of vehicle fleets, cutting down on fuel use and enhancing route planning.
Retail	**Store Layout Optimization:** Retailers employ digital twins to improve customer experiences and sales by optimizing store layouts and product placements.
Pharmaceuticals and Healthcare	**Drug Manufacturing:** To provide exact control over the manufacturing processes and product quality, digital twins are utilized in the pharmaceutical manufacturing industry.

TOOLS FOR DIGITAL TWIN TECHNOLOGY FOR INDUSTRIAL AUTOMATION

To efficiently develop, maintain, and use digital twins for industrial automation, a mix of hardware and software technologies is needed (Table 12.3).

The precise requirements and objectives of the digital twin project, the complexity of the industrial automation system, and the resources at hand all play a role in choosing the best combination of these tools. A well-planned and integrated ecosystem of tools and technologies is necessary for the successful implementation of digital twin technology.

LIST OF INDUSTRIES APPLYING DIGITAL TWIN TECHNOLOGY FOR INDUSTRIAL AUTOMATION

Various sectors have embraced digital twin technology for industrial automation and optimization. Here is a list of sectors using digital twin technology.

- **Manufacturing:** Digital twins are used in manufacturing facilities to enhance product quality, monitor equipment health, and optimize production processes.
- **Energy and Utilities:** Digital twins are used by power plants, renewable energy facilities, and utility corporations for energy optimization, grid management, and preventive maintenance.
- **Aerospace and Defense:** Digital twins are used by the aerospace industry to optimize maintenance schedules, monitor the health of aircraft, and enhance the design and performance of aircraft components.
- **Automotive:** Digital twins are used in the automobile sector for quality assurance, vehicle design, and production process optimization.
- **Oil and Gas:** Oil and gas businesses use digital twins to manage offshore and onshore installations, check the condition of equipment, and improve drilling operations.
- **Pharmaceuticals and Healthcare:** Pharmaceutical companies use digital twin technology to streamline drug production procedures and guarantee product quality.
- **Mining and Natural Resources:** Digital twins are used by mining businesses to monitor mining operations, maximize equipment utilization, and improve safety precautions.
- **Agriculture:** Digital twins are used in precision agriculture to track crop health, better allocate resources, and increase yields.
- **Transportation and Logistics:** Digital twins are used by logistics service providers and transportation firms to manage vehicle fleets, optimize routes, and enhance supply chain operations.
- **Building and Construction:** Building management, energy efficiency, and facility maintenance are all aided by digital twins of the structures [12, 13].

Table 12.3 Common crucial devices for digital twin technology

Device	Description	Usage
IoT Sensors and Devices	IoT sensors and gadgets are placed in the real world to gather information from machinery, processes, and equipment.	For the purpose of constructing digital twins, these sensors collect real-time data on elements including temperature, pressure, humidity, vibration, and energy usage.
Data Acquisition Systems	IoT sensor data is collected, processed, and transmitted via data-gathering systems to digital twin platforms.	These systems make sure that sensor data is reliably recorded and transmitted to the digital twin platform for additional analysis.
Digital Twin Platforms	Software programs called "digital twin platforms" make it possible to create, view, and manage digital twins.	These systems enable engineers and operators to build digital twins and communicate with them, analyze data, and model scenarios for decision-making and optimization.
Data Analytics and Machine Learning Tools	The enormous amounts of data gathered from digital twins and real-world systems are processed and analyzed using data analytics and machine learning methods.	In order to provide predictive maintenance, process optimization, and data-driven insights, these technologies assist in spotting patterns, anomalies, and trends in data.
Simulation and Modeling Software	Digital twins are created using simulation and modeling tools to mimic physical systems virtually.	These technologies help engineers create precise models of the tools, procedures, and environments so they may simulate and test numerous situations and configurations.
Cloud Computing and Edge Computing Platforms	Platforms for cloud and edge computing offer the processing power and storage required to process and store data from digital twins.	Edge computing systems enable real-time analysis and decision-making at the network's edge, whereas cloud platforms are utilized for scalable data storage and processing.

(Continued)

Table 12.3 (Continued) Common crucial devices for digital twin technology

Device	Description	Usage
Visualization and 3D Modeling Tools	Digital twins are created with the use of visualization tools and 3D modeling software.	These tools enable 3D visualization and interaction with digital twins, which improves comprehension and decision-making.
Cybersecurity Solutions	Protecting digital twin data and infrastructure from online threats and unlawful access requires cybersecurity solutions.	To protect the security of digital twin systems, these solutions include firewalls, encryption, access controls, and intrusion detection systems.
Communication Protocols	Data transmission between IoT devices, data collecting systems, and digital twin platforms depends on communication protocols.	In context with digital twins, seamless data sharing and communication are made possible by standardized protocols like MQTT, OPC UA, and CoAP.
Database and Data Storage Systems	For the purpose of storing historical data gathered from digital twins, reliable database and data storage solutions are essential.	These systems give historical data a dependable and expandable store, enabling trend analysis and long-term insights.
Collaboration and Project Management Tools	Teams can plan and manage digital twin projects with the use of collaboration and project management technologies.	To guarantee efficient project execution, teams use these technologies for task management, document sharing, and communication.
Integration and Middleware Solutions	Sensors, data platforms, and control systems are just a few of the elements of the digital twin ecosystem that can be connected with the aid of integration and middleware solutions.	These solutions guarantee smooth data transfer and device interoperability across many technologies.

- **Water and Wastewater Management:** Water treatment facilities and wastewater management systems are monitored and optimized using digital twins.
- **Retail:** Retailers use digital twins to enhance consumer experiences, inventory management, and store layouts.
- **Chemical and Process Industries:** Digital twins are used by businesses in the chemical and process sectors to streamline complicated manufacturing procedures and guarantee product quality.
- **Smart Cities:** To enhance urban planning, infrastructure management, and sustainability efforts, digital twin technology is utilized to produce digital reproductions of entire cities [17].
- **Food and Beverage:** Digital twins are used in the food and beverage sector for process improvement, quality assurance, and inventory management.
- **Maritime and Shipping:** Digital twins are used by the marine sector to monitor vessel health, optimize routes, and guarantee that ships and ports operate effectively.
- **Renewable Energy:** To maximize energy production, renewable energy suppliers utilize digital twins to monitor wind farms, solar installations, and energy storage systems.
- **Chemical Engineering:** Chemical engineering uses digital twins for process improvement, safety analysis, and product development.
- **Metals and Mining:** Digital twins are used in the metals and mining sector for equipment monitoring, production optimization, and preventive maintenance.
- **Pulp and Paper:** The use of digital twins helps to reduce waste, increase resource efficiency, and optimize pulp and paper manufacturing operations.

These sectors have realized the benefits of using digital twin technology to boost productivity, cut costs, enhance product quality, and guarantee the dependability and safety of their operations. Applications [18] for the technology are anticipated to grow as it develops across a variety of industrial industries.

USE CASES OF DIGITAL TWIN TECHNOLOGY FOR INDUSTRIAL AUTOMATION

A variety of industries are using digital twin technology for industrial automation, and there are many applications for it (Table 12.4).

The adaptability and revolutionary possibilities of digital twin technology in industrial automation are demonstrated by these application examples. Organizations can increase efficiency, lower costs, and improve the dependability and sustainability of their operations by developing digital duplicates of their physical systems and processes.

Table 12.4 Notable use cases for digital twins

Use case	Benefits
Predictive Maintenance	Digital twins are used in manufacturing facilities, power plants, and transportation businesses to forecast equipment breakdowns. Before serious problems arise, maintenance alerts are sent out as a result of real-time data collected by sensors and suspicious patterns identified by analytics algorithms.	Decreased downtime, cheaper maintenance, and prolonged equipment life.
Manufacturing Process Optimization	The creation of digital twins extends to entire production lines. Data in real time is gathered from processes and machines. Engineers maximize manufacturing efficiency, cut waste, and improve product quality using simulations and analytics.	Increased product quality, decreased energy use, and increased production yield.
Energy Management	Digital twins are used in smart grids, factories, and buildings to track and manage energy use. Sensors gather information about the equipment, HVAC, and lighting. Energy-saving suggestions are made by algorithms.	Less expensive energy, a smaller carbon footprint, and more sustainability.
Smart Cities	To simulate urban infrastructure, traffic patterns, and public services, smart cities build digital twins. Traffic management and resource allocation are made possible by real-time data from sensors and cameras.	Enhanced public services, lessened congestion, and better urban planning.
Aerospace and Aircraft Maintenance	Digital twins are used by airlines and aerospace firms to track the health of their aircraft. Engine and component sensors gather data, enabling predictive maintenance to stop breakdowns midflight.	Increased airplane security, lower maintenance expenses, and less flight delays.
Healthcare	To monitor and improve the performance of vital medical equipment like MRI machines, hospitals utilize digital twins. Equipment availability and dependability are guaranteed by real-time data and predictive analytics.	Better patient care, less downtime, and financial savings.

(Continued)

Table 12.4 (Continued) Notable use cases for digital twins

	Use case	Benefits
Supply Chain Management	Businesses develop digital twins of their supply chains, including the inventory, warehouses, and transportation. Logistics, demand forecasts, and inventory levels are all optimized using real-time data and simulations.	Better on-time delivery, decreased operational expenses, and effective resource use.
Mining Operations	Digital twins are used by mining businesses to monitor and improve many aspects of their operations, including safety, ore grades, and equipment health. Simulated data and real-time information are used to efficiently plan and carry out mining operations.	Greater productivity, reduced environmental impact, and improved safety.
Water and Wastewater Management	Digital twins are used by utilities and water treatment facilities to track and improve the distribution, sewage treatment, and water quality processes. Water safety and resource efficiency are guaranteed by real-time data.	Greater environmental compliance, lower operating costs, and better water quality.
Oil and Gas Industry	Digital twins are used by the oil and gas industry to monitor and improve pipelines, offshore platforms, and drilling operations. Equipment failures and oil spills are prevented using real-time data and predictive analytics.	Enhanced environmental stewardship, decreased downtime, and increased safety.

REFERENCES

1. Gehrmann, C., & Gunnarsson, M. (2019). A digital twin based industrial automation and control system security architecture. *IEEE Transactions on Industrial Informatics*, 16(1), 669–680.
2. Löcklin, A., Müller, M., Jung, T., Jazdi, N., White, D., & Weyrich, M. (2020, September). Digital twin for verification and validation of industrial automation systems–A survey. In *2020 25th IEEE International Conference on Emerging Technologies and Factory Automation (ETFA)* (Vol. 1, pp. 851–858). IEEE.
3. Cortés, D., Ramírez, J., Villagómez, L., Batres, R., Vasquez-Lopez, V., & Molina, A. (2020, June). Digital Pyramid: An approach to relate industrial automation and digital twin concepts. In *2020 IEEE International Conference on Engineering, Technology and Innovation (ICE/ITMC)* (pp. 1–7). IEEE.
4. Jazdi, N., Talkhestani, B. A., Maschler, B., & Weyrich, M. (2021). Realization of AI-enhanced industrial automation systems using intelligent Digital Twins. *Procedia CIRP*, 97, 396–400.
5. Guerra-Zubiaga, D., Kuts, V., Mahmood, K., Bondar, A., Nasajpour-Esfahani, N., & Otto, T. (2021). An approach to develop a digital twin for industry 4.0 systems: manufacturing automation case studies. *International Journal of Computer Integrated Manufacturing*, 34(9), 933–949.
6. Huang, H., Yang, L., Wang, Y., Xu, X., & Lu, Y. (2021). Digital twin-driven online anomaly detection for an automation system based on edge intelligence. *Journal of Manufacturing Systems*, 59, 138–150.
7. Assawaarayakul, C., Srisawat, W., Ayuthaya, S. D. N., & Wattanasirichaigoon, S. (2019, December). Integrate digital twin to exist production system for industry 4.0. In *2019 4th Technology Innovation Management and Engineering Science International Conference (TIMES-iCON)* (pp. 1–5). IEEE.
8. Redelinghuys, A. J. H., Basson, A. H., & Kruger, K. (2020). A six-layer architecture for the digital twin: A manufacturing case study implementation. *Journal of Intelligent Manufacturing*, 31, 1383–1402.
9. Agrawal, A., Thiel, R., Jain, P., Singh, V., & Fischer, M. (2023). Digital Twin: Where do humans fit in? *Automation in Construction*, 148, 104749.
10. Ghasemi, G., Müller, M., Jazdi, N., & Weyrich, M. (2023). Complexity estimation service for change management in industrial automation systems using Digital Twin. *Procedia CIRP*, 119, 1011–1016.
11. Siddiqui, M., Kahandawa, G., & Hewawasam, H. S. (2023, March). Artificial intelligence enabled digital twin for predictive maintenance in industrial automation system: A novel framework and case study. In *2023 IEEE International Conference on Mechatronics (ICM)* (pp. 1–6). IEEE.
12. Sasikumar, A., Vairavasundaram, S., Kotecha, K., Indragandhi, V., Ravi, L., Selvachandran, G., & Abraham, A. (2023). Blockchain-based trust mechanism for digital twin empowered industrial internet of things. *Future Generation Computer Systems*, 141, 16–27.
13. Kor, M., Yitmen, I., & Alizadehsalehi, S. (2023). An investigation for integration of deep learning and digital twins towards construction 4.0. *Smart and Sustainable Built Environment*, 12(3), 461–487.

14. Attaran, M., & Celik, B. G. (2023). Digital twin: Benefits, use cases, challenges, and opportunities. *Decision Analytics Journal*, 6(80),100165.
15. Masi, M., Sellitto, G. P., Aranha, H., & Pavleska, T. (2023). Securing critical infrastructures with a cybersecurity digital twin. *Software and Systems Modeling*, 22(2), 689–707.
16. Pinheiro, J., Pinto, R., Gonçalves, G., & Ribeiro, A. (2023, July). Lean 4.0: A digital twin approach for automated cycle time collection and Yamazumi analysis. In *2023 3rd International Conference on Electrical, Computer, Communications and Mechatronics Engineering (ICECCME)* (pp. 1–6). IEEE.
17. Wang, H., Chen, X., Jia, F., & Cheng, X. (2023). Digital twin-supported smart city: Status, challenges and future research directions. *Expert Systems with Applications*, 217, 119531.
18. Xu, H., Wu, J., Pan, Q., Guan, X., & Guizani, M. (2023). A survey on digital twin for industrial internet of things: Applications, technologies and tools. *IEEE Communications Surveys & Tutorials*, 25, pp. 1–8.
19. Tao, F., Zhang, M., & Nee, A. Y. C. (2019). *Digital twin driven smart manufacturing*. Academic Press.

Chapter 13

A theoretical analysis of simple retrieval engine

Vipul Narayan, Swapnita Srivastava, Pawan Kumar Mall, Vimal Kumar, and Shashank Awasthi

INTRODUCTION

These overhauls can plainly affect both the creation and the qualities of a wide degree of things and relationships, with major repercussions for adequacy, work, and conflict. Regardless, in any case, tremendous as these impacts may probably be, modernized speculation in like way can change the improvement cycle itself, with results that might be in like way gigantic, and which may, over the long haul, come to overwhelm the brief impact. Consider the example of Atom sharp, a startup firm that is making novel developments for seeing potential medication competitors (and bug showers) by utilizing neural relationships to expect the bioactivity of up-and-comer particles. The affiliation reports that its monstrous convolutional neural relationship "far beat" the introduction of standard "getting" figuring's. After sensible expecting goliath degrees of information, the affiliation's Atom Net thing is depicted as having the choice to "see" central game plan squares of brand name science and is ready for making fundamentally careful figures of the conceded results of considerable real examinations. Such forward skips hold out the opportunity for enormous updates in the advantage of beginning stage drug screening [1]. Verifiably, Atom insightful development (and that of different affiliations utilizing robotized thinking to move drug revelation or clinical choice) is still at the beginning stage: in any case, their principal results send an impression of being attractive, and no new game plans have genuinely come to convey utilizing these new methodologies. In any case, regardless of whether Atom smart passes on absolutely on its demand, its improvement is illustrative of the expected endeavor to build up another advancement "playbook", one that uses enormous datasets and learning figuring's to partake in clear suspicion for ordinary miracles to energize strategy persuading interventions [2]. Atom sharp, for instance, is directly sending this way to deal with the revelation and improvement of new pesticides and specialists for controlling yield ailments. Particle brilliant model depicts two of the paths drives in man-had thinking can impact improvement. Regardless, at any rate, the beginnings of man-made consideration are totally in the field of programming, and its fundamental

DOI: 10.1201/9781003479031-13

business applications have been in consistently close locale like advanced mechanics, the learning examinations that are directly as of now being made recommend that man-made astuteness may at long last have applications across a wide reach. As indicated by the viewpoint of the money-related issue of progress, there is a principal bundle between the issue of furnishing improvement motivations to make impels with a meekly little space of use, such robots reason worked for restricted undertakings, versus pushes with a wide—assistants may say basically amazing—territory of utilization, as might be legitimate for the advances in neural affiliations and AI regularly recommended as "huge learning." As such, a first deals to be introduced is how much updates in man-made consideration are events of new advances, at any rate rather might be such "all around huge unanticipated turns of events" (later on GPTs) that have genuinely been such engaging drivers of widened length mechanical headway. Second, a couple of places of motorized thinking will unequivocally set up more moderate or extra stunning obligations to many existing creation measures (hitting worries about the potential for colossal occupation clearings), others, as fundamental learning, hold out the opportunity of not just profitability gains across a wide assortment of areas yet also changes in the credible considered the headway relationship inside those spaces. As passed on inside and out, by drawing in progression across different applications, the "arrangement of a met" [3]. We have seen that there ought to be some unaided AI calculation equipped for recognizing substances in a surge of information and making invariant portrayals of them under genuinely powerless suspicions. How should such a calculation resemble this? To recognize something as an element, the calculation needs rehashed comparative perceptions. The reiteration is important, since else it may very well get some clamor for a substance. The issue is the idea of closeness (or rather the deficiency in that department): to choose whether two information focuses are comparative, the calculation needs a measurement of the info information, for example a numerical capacity that decides a distance measure between any two potential info information focuses. At first, it very well may be enticing to settle on a credulous decision like a Euclidean measurement of the information space. In any case, by and by this endeavor doesn't get us far; for instance, on account of picture acknowledgment, a basic interpretation brings about two pictures that are for all intents and purposes indistinguishable (from the natural eye) yet are isolated by a genuinely huge distance in the pixel-by-pixel Euclidean measurement. A valuable measurement would need to consider that two pictures can be completely extraordinary on a pixel-by-pixel premise yet at the same time show a similar article (seen from two alternate points of view, for instance) [4]. Along these lines, in the event that we need to restrict ourselves to unaided learning dependent on just the three suppositions from the past area, at that point we need a calculation that can gain proficiency with the correct measurement in a solo manner from the information. The third presumption (steadiness on schedule) makes that conceivable on a basic level: the measurement distance between two ensuing information focuses ought to normally be little since two resulting information focuses typically are indications of a similar substance (or set of elements). Our calculation should develop and continuously improve a particular metric dependent on the information it gets. There is an issue: the space of potential measurements is cosmically huge and we have just a similarly modest number of hence noticed information focuses. How could the calculation extrapolate from the perceptions to the full space?

LITERATURE SURVEY

As opposed to attempting to clarify what the cerebrum does, we center around how a mind should deal with the abilities that it is known to have [4]. The errand that we have picked is unmistakably not a simple one and backing from or coordinated effort with others is generally welcome. We expect to mirror the interdisciplinary idea of our work behind the scenes of our examination gathering's perpetual staff - covering mastery in neuroscience, software engineering, arithmetic, and hypothetical physical science [5]. What's more, we are keen on developing joint efforts with scholastic exploration bunches from the referenced fields. Given our aspiration to reevaluate the fundamentals of AI, we additionally feel that significant commitments may come from splendid junior analysts with restricted insight (and inclination). We are subsequently effectively captivating with understudies and youthful specialists through entry-level positions, college rivalries, and comparative configurations, attempting to produce whatever number of new thoughts as could be allowed for this interesting and significant field of science. Particle smart movement (and that of various affiliations using robotized thinking to move drug disclosure or clinical decision) is still at an early phase: regardless of their fundamental results transmit an impression of being drawing in, no new arrangements have truly come to communicate using these new philosophies [6]. Regardless, whether or not Atom wise passes on totally on its assertion, its improvement is illustrative of the anticipated undertaking to develop another development "playbook", one that utilizes gigantic datasets and learning figuring's to take part in undeniable assumption for ordinary supernatural occurrences to encourage strategy convincing interventions. Molecule sharp, for example, is right now sending this approach to manage the disclosure and development of new pesticides and experts for controlling yield ailments. Particle astute model portrays two of the ways pushes in man-had thinking can affect improvement [7]. Notwithstanding, at any rate, the beginnings of man-made care are completely in the field of programming, and its essential business applications have been in ordinarily close locales like progressed mechanics, the learning evaluations that are right as of now being made recommend that man-made acumen may finally have applications across a wide reach.

METHODOLOGY

Here is a harsh diagram of how such a calculation may function:

1. **Recognize Basic Substances:** toward the start of our learning cycle, it is unmistakably difficult to perceive complex elements (like pictures of a feline or a full tune), since normally there will not be rehashed indistinguishable perceptions of a similar element and we don't yet have a valuable idea of likeness. It is conceivable however to perceive certain rudimentary substances dependent on insights; for instance, in pictures, the calculation would discover edges as rehashing designs. In addition, on account of music, it may recognize certain mixes of frequencies (for example tones and hints) as rudimentary examples.

2. **Learn Changes:** by noticing the time reliance of the identified rudimentary substances, essential transformations could be learned. For instance, on account of video arrangements, the calculation would discover that edges will in general move in specific ways that are exceptionally associated.

3. **Refine the Measurement:** in view of the learned changes, certain arrangements of rudimentary elements could be assembled in equivalency classes. For instance, the calculation may conclude that a bunch of edges that show up in specific positions compared with one another are consistently "something very similar", autonomous of where this arrangement of edges is situated in the picture. (Placing mixes of elements into equivalency classes basically refines the measurement, for example, the idea of closeness, on the info space: objects in a similar class should be extremely near one another in that measurement sense.)

4. **Recognize More Perplexing Elements:** Thanks to the refined measurement, new bunches of information focus arise and can be distinguished through unaided learning. The information focuses that gave off an impression of being far separated in our underlying innocent measurement may really end up being important for a similar equivalency class. Therefore, the calculation can learn elements that are more mind-boggling than those found in the initial step.

5. **Iterate:** more unpredictable elements will display more perplexing changes, which thus consider bigger equivalency classes, which at that point empower the calculation to recognize much more elevated level substances.

6. **Learn Successions:** the solo substance recognition measure depicted above should be entwined with the learning of rehashed arrangements on the various degrees of deliberation. For instance, the sound of a solitary piano key is a succession of recurrence designs. An entire tune, while likewise a succession, should be addressed fair and square of notes and spans to be free from the instrument on which it is played. Plainly the idea above still requires a decent lot of explanation and

itemizing. A more straightforward form of it has been proposed and executed by Jeff Hawkins' organization Momenta under the name Hierarchical Temporal Memory (HTM) [8]. The principal expansion of our methodology contrasted with Numina's HTM is the plan to utilize the transient development of the info sign to learn changes, make equivalency classes from that, and consequently open more significant level elements for solo learning. The essential target of our exploration is to make new calculations that dive past deep figuring out how to perform intellectual assignments in a manner that is like how natural cerebrums think. Considering every one of the thoughts and contemplations from this white paper, we have characterized the accompanying examination procedure to seek after this objective: Zero in on invariant portrayals: the way to Strong AI is a hypothesis of the mind – the cerebrum's center layer. The way into the mind's center layer are invariant portrayals. Unaided and presumption poor: in the mind, invariant portrayals appear to be framed in an unsuppler-vised path by a solitary cortical calculation. It is thusly our need to discover calculations that make invariant portrayals without management and with negligible presumptions on the info information. Reevaluate the nuts and bolts: we work to comprehend the establishments of knowledge and to discover fundamentally better approaches for making AI. It isn't our essential goal to improve characterization precision in some AI challenges by one more, not many rates focus. This implies that we will now and then pick levels of reflection that may restrict the common convenience in the present moment yet that may empower us more readily to get knowledge.

- **Interdisciplinary Approach**: we expect to take care of the issue of invariant portrayals from three points:
    1. **The Theoretical Methodology**: characterize numerical articles that depict general substances and fabricate a "hypothesis of invariant portrayals" on top,
    2. **The Programming Approach**: fabricate model calculations that execute parts of the hypothesis, and
    3. **The Neuroscience Approach**: recognize neural circuits that execute parts of the hypothesis of invariant portrayals to approve and change the hypothesis and to improve our comprehension of the mind. We accept that to comprehend the mind, you need to develop it.

RESULT

This article begins to debilitate the conceivable impact of advances in man-gained allowance on headway and to see the work that system and foundations may play in giving stunning inspirations to advance, scattering, and contention around there. We start in Section 13.2 by including the specific cash-related issue of assessment contraptions, of which basic learning applied to R&D issues is an especially energizing model. We turn around the trade between the degree of mutilation of use of another assessment instrument and the piece of assessment gadgets not simply in improving the efficiency of evaluation improvement but in making another "playbook" for development itself. We by then turn in Section 13.3 to quickly disconnect three key inventive heading inside AI—mechanical headway, huge developments, and tremendous learning. We recommend that as much of the time as possible conflated fields will probably recognize totally amazing parts later on for development and explicit change (Figures 13.1 and 13.2).

Work in basic plans appears to have dropped down and is certainly going to have tolerably little impact going advances. Moreover, reviewing those developments in mechanical headway can also get human work in the creation free from various things and undertakings, improvement in bleeding edge mechanics moves essentially has normally low potential to change the opportunity of progress itself. Obviously, colossal learning is plainly an area of assessment that is especially totally strong and that can change the progress correspondence itself (Figure 13.3).

CONCLUSION

According to the perspective of the cash-related issue of progress (among others, Bresnahan and Trajtenberg (1995)), there is an essential parcel between the issue of outfitting improvement inspirations to make propels with a submissively little space of utilization, such robots reason worked for limited endeavors, versus pushes with a wide—accomplices may say fundamentally incredible—area of usage, as may be valid for the advances in neural affiliations and AI consistently suggested as "enormous learning." As such, a first request to be presented is how much updates in man-made care are occasions of new advances, at any rate rather may be such "all around significant unforeseen developments" (later on GPTs) that have truly been such appealing drivers of broadened length mechanical advancement. Second, a few positions of mechanized reasoning will emphatically set up more moderate or extra shocking duties to many existing creation measures (jabbing stresses over the potential for goliath occupation clearings), while others, as basic learning, hold out the chance of not simply benefit gains across a wide variety of districts yet similarly changes in the authentic contemplated the advancement relationship inside those spaces. As passed on altogether by Gribiche (1957), by enabling advancement across various applications, the "formation of a technique for progress" can have basically more essential cash-related impact than the progress of any single new thing. Here we battle that new advances in AI and neural relationship, through their ability to improve both the exhibit of end-use propels and

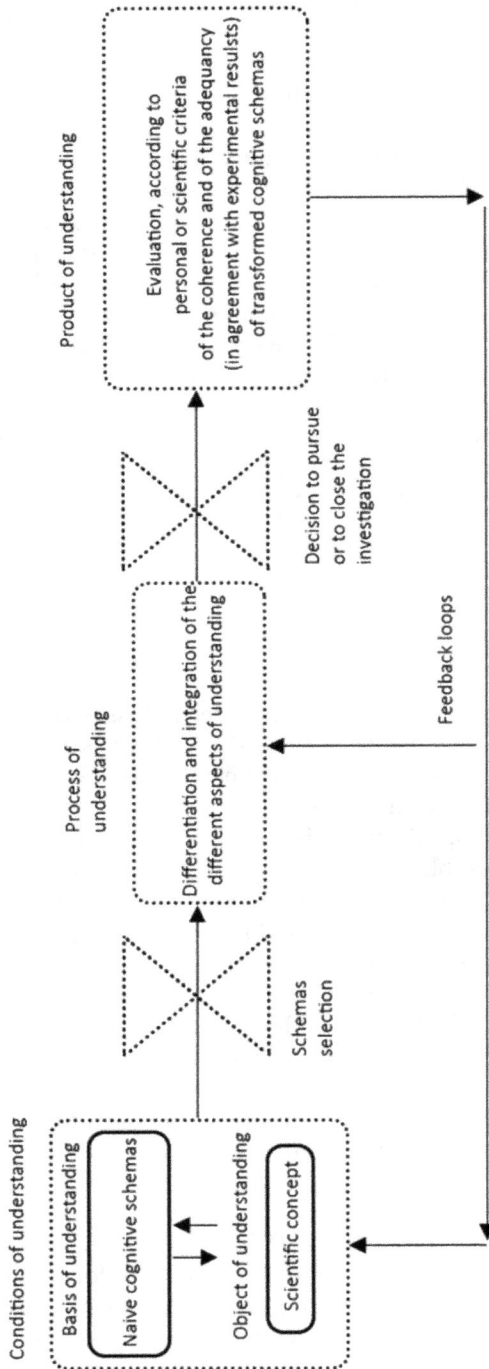

Figure 13.1 Different model determination for different schemas.

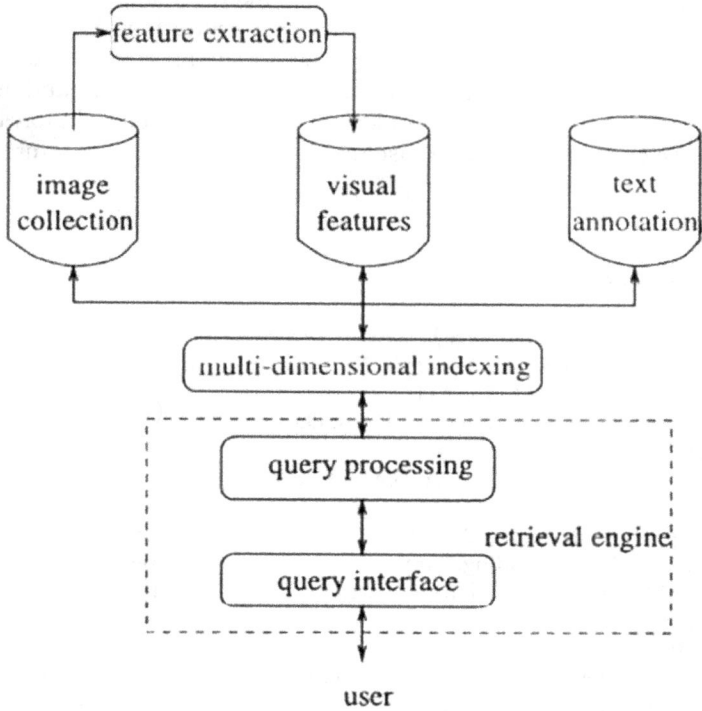

Figure 13.2 Complete flow of data.

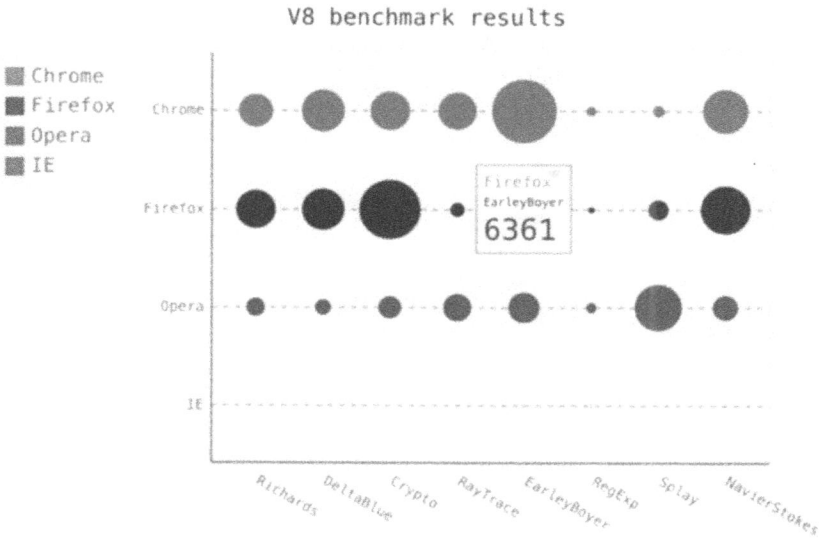

Figure 13.3 VB benchmark visualization.

the possibility of the progress affiliation, are likely passing all-around sway movement and improvement [8]. As such the inspirations and blocks that may shape the unforeseen new development and spread of these kinds of progress are an enormous subject for cash-related assessment, and building a perspective on the conditions under which certain potential pioneers can get to these contraptions and to use them in a valuable for guaranteed way is a central concern for strategy.

REFERENCES

[1] M. Bedny et al.: "Language Processing in the Occipital Cortex of Congenitally Blind Adults." *Proceedings of the National Academy of Sciences* 108, no. 11 (March 15, 2011) 4429–4434.

[2] N. Bostrom: *Superintelligence - Paths, Dangers, Strategies* (Oxford University Press, 2014).

[3] E. Brynjolfsson, A. McAfee: *The Second Machine Age* (Norton Paperback, 2016).

[4] J. Devlin, et al.: "Bert: Pre-Training of Deep Bidirectional Transformers for Language Understanding." arXiv preprint arXiv:1810.04805 (2018).

[5] J. J. DiCarlo, D. Zoccolan, N. C. Rust: "How Does the Brain Solve Visual Object Recognition?" *Neuron* 73, no. 3 (February 9, 2012) 415–434. doi:10.1016/j.neuron.2012.01.010.

[6] B. Fritzke: "A growing neural gas network learns topologies." In: Tesauro, G.; Touretzky, D. S. and Leen, T. K. (Eds.), *Advances in Neural Information Processing Systems 7*, MIT Press, 1995, 625–632.

[7] Future of Life Institute: "An Open Letter - Research Priorities for Robust and Beneficial Artificial Intelligence." https://futureoflife.org/ai-open-letter/?cn-reloaded=1.

[8] J. Hawkins, S. Ahmad: "Why Neurons Have Thousands of Synapses: A Theory of Sequence Memory in Neocortex." *Frontiers in Neural Circuits* 10 (2016) 1–13.

Index

For Product Safety Concerns and Information please contact our EU
representative GPSR@taylorandfrancis.com
Taylor & Francis Verlag GmbH, Kaufingerstraße 24, 80331 München, Germany

www.ingramcontent.com/pod-product-compliance
Lightning Source LLC
Chambersburg PA
CBHW060353220326
41598CB00023B/2905

9 781032 765570